blogdown
Creating Websites
with R Markdown

Chapman & Hall/CRC
The R Series

Series Editors

John M. Chambers
Department of Statistics
Stanford University
Stanford, California, USA

Torsten Hothorn
Division of Biostatistics
University of Zurich
Switzerland

Duncan Temple Lang
Department of Statistics
University of California, Davis
Davis, California, USA

Hadley Wickham
RStudio
Boston, Massachusetts, USA

Aims and Scope

This book series reflects the recent rapid growth in the development and application of R, the programming language and software environment for statistical computing and graphics. R is now widely used in academic research, education, and industry. It is constantly growing, with new versions of the core software released regularly and more than 10,000 packages available. It is difficult for the documentation to keep pace with the expansion of the software, and this vital book series provides a forum for the publication of books covering many aspects of the development and application of R.

The scope of the series is wide, covering three main threads:
- Applications of R to specific disciplines such as biology, epidemiology, genetics, engineering, finance, and the social sciences.
- Using R for the study of topics of statistical methodology, such as linear and mixed modeling, time series, Bayesian methods, and missing data.
- The development of R, including programming, building packages, and graphics.

The books will appeal to programmers and developers of R software, as well as applied statisticians and data analysts in many fields. The books will feature detailed worked examples and R code fully integrated into the text, ensuring their usefulness to researchers, practitioners and students.

Published Titles

Stated Preference Methods Using R, *Hideo Aizaki, Tomoaki Nakatani, and Kazuo Sato*

Using R for Numerical Analysis in Science and Engineering, *Victor A. Bloomfield*

Event History Analysis with R, *Göran Broström*

Extending R, *John M. Chambers*

Computational Actuarial Science with R, *Arthur Charpentier*

Testing R Code, *Richard Cotton*

The R Primer, Second Edition, *Claus Thorn Ekstrøm*

Statistical Computing in C++ and R, *Randall L. Eubank and Ana Kupresanin*

Basics of Matrix Algebra for Statistics with R, *Nick Fieller*

Reproducible Research with R and RStudio, Second Edition, *Christopher Gandrud*

R and MATLAB® *David E. Hiebeler*

Statistics in Toxicology Using R *Ludwig A. Hothorn*

Nonparametric Statistical Methods Using R, *John Kloke and Joseph McKean*

Displaying Time Series, Spatial, and Space-Time Data with R, *Oscar Perpiñán Lamigueiro*

Programming Graphical User Interfaces with R, *Michael F. Lawrence and John Verzani*

Analyzing Sensory Data with R, *Sébastien Lê and Theirry Worch*

Parallel Computing for Data Science: With Examples in R, C++ and CUDA, *Norman Matloff*

Analyzing Baseball Data with R, *Max Marchi and Jim Albert*

Growth Curve Analysis and Visualization Using R, *Daniel Mirman*

R Graphics, Second Edition, *Paul Murrell*

Introductory Fisheries Analyses with R, *Derek H. Ogle*

Data Science in R: A Case Studies Approach to Computational Reasoning and Problem Solving, *Deborah Nolan and Duncan Temple Lang*

Multiple Factor Analysis by Example Using R, *Jérôme Pagès*

Customer and Business Analytics: Applied Data Mining for Business Decision Making Using R, *Daniel S. Putler and Robert E. Krider*

Flexible Regression and Smoothing: Using GAMLSS in R, *Mikis D. Stasinopoulos, Robert A. Rigby, Gillian Z. Heller, Vlasios Voudouris, and Fernanda De Bastiani*

Implementing Reproducible Research, *Victoria Stodden, Friedrich Leisch, and Roger D. Peng*

Published Titles continued

Graphical Data Analysis with R, *Antony Unwin*

Using R for Introductory Statistics, Second Edition, *John Verzani*

Advanced R, *Hadley Wickham*

bookdown: Authoring Books and Technical Documents with R Markdown, *Yihui Xie*

blogdown: Creating Websites with R Markdown, *Yihui Xie, Amber Thomas, and Alison Presmanes Hill*

Dynamic Documents with R and knitr, Second Edition, *Yihui Xie*

blogdown
Creating Websites with R Markdown

Yihui Xie
Amber Thomas
Alison Presmanes Hill

CRC Press
Taylor & Francis Group
Boca Raton   London   New York

CRC Press is an imprint of the
Taylor & Francis Group, an **informa** business

A CHAPMAN & HALL BOOK

CRC Press
Taylor & Francis Group
6000 Broken Sound Parkway NW, Suite 300
Boca Raton, FL 33487-2742

© 2018 by Taylor & Francis Group, LLC
CRC Press is an imprint of Taylor & Francis Group, an Informa business

No claim to original U.S. Government works

Printed on acid-free paper

International Standard Book Number-13: 978-0-8153-6372-9 (Paperback)
International Standard Book Number-13: 978-0-8153-6384-2 (Hardback)

Visit the Taylor & Francis Web site at
http://www.taylorandfrancis.com

and the CRC Press Web site at
http://www.crcpress.com

百千万劫弹指过，春夏秋心凭谁托。
日月星灭观夜落，天地人间对文酌。

Life is short. Write for eternity.

Contents

D Advanced Topics   **123**

E Personal Experience   **139**

Bibliography   **141**

Index   **143**

List of Tables

List of Figures

Preface

In the summer of 2012, I did my internship at AT&T Labs Research,[1] where I attended a talk given by Carlos Scheidegger (`https://cscheid.net`), and Carlos said something along the lines of "if you don't have a website nowadays, you don't exist." Later I paraphrased it as:

"I web, therefore I am ~~a spiderman~~."

Carlos's words resonated very well with me, although they were a little exaggerated. A well-designed and maintained website can be extremely helpful for other people to know you, and you do not need to wait for suitable chances at conferences or other occasions to introduce yourself in person to other people. On the other hand, a website is also highly useful for yourself to keep track of what you have done and thought. Sometimes you may go back to a certain old post of yours to relearn the tricks or methods you once mastered in the past but have forgotten.

We introduce an R package, **blogdown**, in this short book, to teach you how to create websites using R Markdown and Hugo. If you have experience with creating websites, you may naturally ask what the benefits of using R Markdown are, and how **blogdown** is different from existing popular website platforms, such as WordPress. There are two major highlights of **blogdown**:

1.   It produces a static website, meaning the website only consists of static files such as HTML, CSS, JavaScript, and images, etc. You can host the website on any web server (see Chapter 3 for

---

[1] In this book, "I" and "my" refer to Yihui unless otherwise noted.

details). The website does not require server-side scripts such as PHP or databases like WordPress does. It is just one folder of static files. We will explain more benefits of static websites in Chapter 2, when we introduce the static website generator Hugo.

2. The website is generated from R Markdown documents (R is optional, i.e., you can use plain Markdown documents without R code chunks). This brings a huge amount of benefits, especially if your website is related to data analysis or (R) programming. Being able to use Markdown implies simplicity and more importantly, *portability* (e.g., you are giving yourself the chance to convert your blog posts to PDF and publish to journals or even books in the future). R Markdown gives you the benefits of dynamic documents — all your results, such as tables, graphics, and inline values, can be computed and rendered dynamically from R code, hence the results you present on your website are more likely to be reproducible. An additional yet important benefit of using R Markdown is that you will be able to write technical documents easily, due to the fact that **blogdown** inherits the HTML output format from **bookdown** (Xie, 2016). For example, it is possible to write LaTeX math equations, BibTeX citations, and even theorems and proofs if you want.

Please do not be misled by the word "blog" in the package name: **blogdown** is for general-purpose websites, and not only for blogs. For example, all authors of this book have their personal websites, where you can find information about their projects, blogs, package documentations, and so on.[2] All their pages are built from **blogdown** and Hugo.

If you do not prefer using Hugo, there are other options, too. Chapter

---

[2]Yihui's homepage is at `https://yihui.name`. He writes blog posts in both Chinese (`https://yihui.name/cn/`) and English (`https://yihui.name/en/`), and documents his software packages such as **knitr** (`https://yihui.name/knitr/`) and **animation** (`https://yihui.name/animation/`). Occasionally he also writes articles like `https://yihui.name/rlp/` when he finds interesting topics but does not bother with a formal journal submission. Amber's homepage is at `https://amber.rbind.io`, where you can find her blog and project pages. Alison's website is at `https://alison.rbind.io`, which uses an academic theme at the moment.

5 presents possibilities of using other site generators, such as Jekyll and **rmarkdown**'s default site generator.

Structure of the book

Chapter 1 aims at getting you started with a new website based on **blogdown**: it contains an installation guide, a quick example, an introduction to RStudio addins related to **blogdown**, and comparisons of different source document formats. All readers of this book should finish at least this chapter (to know how to create a website locally) and Section 3.1 (to know how to publish a website). The rest of the book is mainly for those who want to further customize their websites.

Chapter 2 briefly introduces the static website generator Hugo, on which **blogdown** is based. We tried to summarize the official Hugo documentation in a short chapter. You should consult the official documentation when in doubt. You may skip Section 2.5 if you do not have basic knowledge of web technologies. However, this section is critical for you to fully understand Hugo. We have spent the most time on this section in this chapter. It is very technical, but should be helpful nonetheless. Once you have learned how to create Hugo templates, you will have the full freedom to customize your website.

Chapter 3 tells you how to publish a website, so that other people can visit it through a link. Chapter 4 shows how to migrate existing websites from other platforms to Hugo and **blogdown**. Chapter 5 gives a few other options if you do not wish to use Hugo as your site generator.

Appendix A is a quick tutorial on R Markdown, the prerequisite of **blogdown** if you are going to write R code in your posts. Appendix B contains basic knowledge about websites, such as HTML, CSS, and JavaScript. If you really care about your website, you will have to learn them someday. If you want to have your own domain name, Appendix C provides an introduction to how it works. We have also covered some optional topics in Appendix D for advanced users.

Software information and conventions

The R session information when compiling this book is shown below:

```
sessionInfo()
```

```
## R version 3.4.2 (2017-09-28)
## Platform: x86_64-apple-darwin15.6.0 (64-bit)
## Running under: macOS Sierra 10.12.6
##
## Matrix products: default
##
## locale:
## [1] en_US.UTF-8/en_US.UTF-8/en_US.UTF-8/C/en_US.UTF-8/en_US.UTF-8
##
## attached base packages:
## [1] stats     graphics  grDevices utils     datasets
## [6] base
##
## loaded via a namespace (and not attached):
## [1] bookdown_0.6    blogdown_0.2    rmarkdown_1.7
## [4] htmltools_0.3.6 knitr_1.18
```

We do not add prompts (> and +) to R source code in this book, and we comment out the text output with two hashes ## by default, as you can see from the R session information above. This is for your convenience when you want to copy and run the code (the text output will be ignored since it is commented out). Package names are in bold text (e.g., **rmarkdown**), and inline code and filenames are formatted in a typewriter font (e.g., knitr::knit('foo.Rmd')). Function names are followed by parentheses (e.g., blogdown::serve_site()). The double-colon operator :: means accessing an object from a package.

A trailing slash often indicates a directory name, e.g., content/ means a directory named content instead of a file named content. A leading slash in a path indicates the root directory of the website, e.g., /static/css/style.css means the file `static/css/style.css` under the root directory of your website project instead of your operating system. Please note that some directory names are configurable, such as `public/`, but we will use their default values throughout the book. For example, your website will be rendered to the `public/` directory by default, and when you see `public/` in this book, you should think of it as the actual publishing directory you set if you have changed the default value. `Rmd` stands for R Markdown in this book, and it is the filename extension of R Markdown files.

A "post" often does not literally mean a blog post, but refers to any source documents (Markdown or R Markdown) in the website project, including blog posts and normal pages. Typically blog posts are stored under the `content/post/` directory, and pages are under other directories (including the root `content/` directory and its subdirectories), but Hugo does not require this structure.

The URL `http://www.example.com` is used only for illustration purposes. We do not mean you should actually visit this website. In most cases, you should replace `www.example.com` with your actual domain name.

An asterisk `*` in a character string often means an arbitrary string. For example, `*.example.com` denotes an arbitrary subdomain of `example.com`. It could be `foo.example.com` or `123.example.com`. Actually, `foo` and `bar` also indicate arbitrary characters or objects.

Acknowledgments

Originally I planned to write only one sentence in this section: "I thank Tareef." This book and the **blogdown** package would not have been finished without Tareef, the president of RStudio. He has been "gently nudging" me every week since Day 1 of **blogdown**. As a person without strong self-discipline and working remotely, I benefited a lot from weekly meetings with him. He also gave me a lot of good technical suggestions on improving the package. Actually, he was one of the very earliest users of **blogdown**.

Of course, I'd like to thank RStudio for the wonderful opportunity to work on this new project. I was even more excited about **blogdown** than **bookdown** (my previous project). I started blogging 12 years ago, and have used and quit several tools for building websites. Finally I feel satisfied with my own dog food.

Many users have provided helpful feedback and bug reports through GitHub issues (https://github.com/rstudio/blogdown/issues). Two of my favorites are https://github.com/rstudio/blogdown/issues/40 and https://github.com/rstudio/blogdown/issues/97. Some users have also contributed code and improved this book through pull requests (https://github.com/rstudio/blogdown/pulls). You can find the list of contributors at https://github.com/rstudio/blogdown/graphs/contributors. Many users followed my suggestion to ask questions on StackOverflow (https://stackoverflow.com/tags/blogdown) instead of using GitHub issues or Emails. I appreciate all your help, patience, and understanding. I also want to make special mention of my little friend Jerry Han, who was probably the youngest **blogdown** user.

For this book, I was fortunate enough to work with my co-authors Amber and Alison, who are exceptionally good at explaining things to beginners. That is the ability I desire most. Needless to say, they have made this book friendlier to beginners. In addition, Sharon Machlis contributed some advice on search engine optimization in this book (https://github.com/rstudio/blogdown/issues/193). Raniere Silva contributed Section 3.5 (https://github.com/rstudio/blogdown/pull/225).

I'd like to thank all Hugo authors and contributors (Bjørn Erik Pedersen and Steve Francia *et al.*) for such a powerful static site generator. At least it made me enjoy building static websites and blogging again.

For some reason, a part of the R community started to adopt the "sticker-driven development" model when developing packages. I was hoping **blogdown** could have a hexbin sticker, too, so I asked for help on Twitter (https://twitter.com/xieyihui/status/907269861574930432) and got tons of draft logos. In particular, I want to thank Thomas Lin Pedersen for his hard work on a very clever design. The final version of the logo was provided by Taras Kaduk and Angelina Kaduk, and I truly appreciate it.

This is the third book I have published with my editor at Chapman & Hall/CRC, John Kimmel. I always love working with him. Rebecca Condit and Suzanne Lassandro proofread the manuscript, and I learned a lot from their comments and professional suggestions.

Yihui Xie
Elkhorn, Nebraska

About the Authors

Yihui is the main developer of the **blogdown** package. He did not start working on the systematic documentation (i.e., this book) until four months after he started the **blogdown** project. One day, he found a very nice **blogdown** tutorial on Twitter written by Amber Thomas. Being surprised that she could create a great personal website using **blogdown** and write a tutorial *when there was no official documentation*, Yihui immediately invited her to join him to write this book, although they had never met each other before. This definitely would not have happened if Amber did not have a website. By the way, Amber asked the very first question[3] with the blogdown tag on StackOverflow.

About half a year later, Yihui noticed another very well-written **blogdown** tutorial by Alison on her personal website, when this book was still not complete. The same story happened, and Alison became the third author of this book. The three authors have never met each other.

Hopefully, you can better see why you should have a website now.

Yihui Xie

Yihui Xie (https://yihui.name) is a software engineer at RStudio (https://www.rstudio.com). He earned his PhD from the Department of Statistics, Iowa State University. He is interested in interactive statistical graphics and statistical computing. As an active R user, he has authored several R packages, such as **knitr**, **bookdown**, **blogdown**, **xaringan**, **animation**, **DT**, **tufte**, **formatR**, **fun**, **mime**, **highr**, **servr**, and **Rd2roxygen**, among which the **ani-**

---

[3] https://stackoverflow.com/q/41176194/559676

mation package won the 2009 John M. Chambers Statistical Software Award (ASA). He also co-authored a few other R packages, including **shiny, rmarkdown**, and **leaflet**.

In 2006, he founded the Capital of Statistics (`https://cosx.org`), which has grown into a large online community on statistics in China. He initiated the Chinese R conference in 2008, and has been involved in organizing R conferences in China since then. During his PhD training at Iowa State University, he won the Vince Sposito Statistical Computing Award (2011) and the Snedecor Award (2012) in the Department of Statistics.

He occasionally rants on Twitter (`https://twitter.com/xieyihui`), and most of the time you can find him on GitHub (`https://github.com/yihui`).

He enjoys spicy food as much as classical Chinese literature.

Amber Thomas

Amber Thomas (`https://amber.rbind.io`) is a data journalist and "maker" at the online publication of visual essays: The Pudding (`https://pudding.cool`). Her educational background, however, was in quite a different field altogether: marine biology. She has a bachelor's degree in marine biology and chemistry from Roger Williams University and a master's degree in marine sciences from the University of New England. Throughout her academic and professional career as a marine biologist, she realized that she had a love of data analysis, visualization, and storytelling and thus, she switched career paths to something a bit more data focused.

While looking for work, she began conducting personal projects to expand her knowledge of R's inner workings. She decided to put all of her projects in a single place online (so that she could be discovered, naturally) and after lots of searching, she stumbled upon an early release of the **blogdown** package. She was hooked right away and spent a few days setting up her personal website and writing a tutorial on how she did it. You can find that tutorial and some of her other projects and musings on her blogdown site.

When she is not crunching numbers and trying to stay on top of her email inbox, Amber is usually getting some fresh Seattle air or cuddling with her dog, Sherlock. If you are looking for her in the digital world, try `https://twitter.com/ProQuesAsker`.

Alison Presmanes Hill

Alison (`https://alison.rbind.io`) is a professor of pediatrics at Oregon Health and Science University's (OHSU) Center for Spoken Language Understanding in Portland, Oregon. Alison earned her PhD in developmental psychology with a concentration in quantitative methods from Vanderbilt University in 2008. Her current research focuses on developing better outcome measures to evaluate the impact of new treatments for children with autism and other neurodevelopmental disorders, using natural language processing and other computational methods. Alison is the author of numerous journal articles and book chapters, and her work has been funded by the National Institutes of Health, the Oregon Clinical and Translational Research Institute, and Autism Speaks.

In addition to research, Alison teaches graduate-level courses in OHSU's Computer Science program (`https://www.ohsu.edu/csee`) on statistics, data science, and data visualization using R. She has also developed and led several R workshops and smaller team-based training sessions, and loves to train new "useRs." You can find some of her workshop and teaching materials on GitHub (`https://github.com/apreshill`) and, of course, on her **blogdown** site.

Being a new mom, Alison's current favorite books are *The Circus Ship* and *Bats at the Ballgame*. She also does rousing renditions of most Emily Arrow songs (for private audiences only).

1

Get Started

In this chapter, we show how to create a simple website from scratch. The website will contain a home page, an "About" page, one R Markdown post, and a plain Markdown post. You will learn the basic concepts for creating websites with **blogdown**. For beginners, we recommend that you get started with the RStudio IDE, but it is not really required. The RStudio IDE can make a few things easier, but you are free to use any editor if you do not care about the extra benefits in RStudio.

1.1 Installation

We assume you have already installed R (https://www.r-project.org) (R Core Team, 2017) and the RStudio IDE (https://www.rstudio.com). If you do not have RStudio IDE installed, please install Pandoc (http://pandoc.org). Next we need to install the **blogdown** package in R. It is available on CRAN and GitHub, and you can install it with:

```r
## Install from CRAN
install.packages("blogdown")
## Or, install from GitHub
if (!requireNamespace("devtools")) install.packages("devtools")
devtools::install_github("rstudio/blogdown")
```

Since **blogdown** is based on the static site generator Hugo (https://gohugo.io), you also need to install Hugo. There is a helper function in **blogdown** to download and install it automatically on major operating systems (Windows, macOS, and Linux):

```
blogdown::install_hugo()
```

By default, it installs the latest version of Hugo, but you can choose a specific version through the version argument if you prefer.

For macOS users, install_hugo() uses the package manager Homebrew (https://brew.sh) if it has already been installed, otherwise it just downloads the Hugo binary directly.

1.1.1 Update

To upgrade or reinstall Hugo, you may use blogdown::update_hugo(), which is equivalent to install_hugo(force = TRUE). You can check the installed Hugo version via blogdown::hugo_version(), and find the latest version of Hugo at https://github.com/gohugoio/hugo/releases.

1.2 A quick example

From our experience, Hugo's documentation may be a little daunting to read and digest for beginners.[1] For example, its "Quickstart" guide used to have 12 steps, and you can easily get lost if you have not used a static website generator before. For **blogdown**, we hope users of all levels can at least get started as quickly as possible. There are many things you may want to tweak for the website later, but the first step is actually fairly simple: create a new project under a new directory in the RStudio IDE (File -> New Project), and call the function in the R console of the new project:

```
blogdown::new_site()
```

Then wait for this function to create a new site, download the default theme,

---

[1]One day I was almost ready to kill myself when I was trying to figure out how _index.md works by reading the documentation over and over again, and desperately searching on the Hugo forum.

add some sample posts, open them, build the site, and launch it in the RStudio Viewer, so you can immediately preview it. If you do not use the RStudio IDE, you need to make sure you are currently in an empty directory,[2] in which case new_site() will do the same thing, but the website will be launched in your web browser instead of the RStudio Viewer.

Now you should see a bunch of directories and files under the RStudio project or your current working directory. Before we explain these new directories and files, let's introduce an important and helpful technology first: *LiveReload*. This means your website[3] will be automatically rebuilt and reloaded in your web browser[4] when you modify any source file of your website and save it. Basically, once you launch the website in a web browser, you do not need to rebuild it explicitly anymore. All you need to do is edit the source files, such as R Markdown documents, and save them. There is no need to click any buttons or run any commands. LiveReload is implemented via blogdown::serve_site(), which is based on the R package **servr** (Xie, 2017d) by default.[5]

The new_site() function has several arguments, and you may check out its R help page (?blogdown::new_site) for details. A minimal default theme named "hugo-lithium-theme" is provided as the default theme of the new site,[6] and you can see what it looks like in Figure 1.1.

You have to know three most basic concepts for a Hugo-based website:

1.   The configuration file config.toml, in which you can specify some global settings for your site. Even if you do not know what TOML is at this point (it will be introduced in Chapter 2), you may still be

---

[2]Check the output of list.files('.') in R, and make sure it does not include files other than LICENSE, the RStudio project file (*.Rproj), README or README.md.

[3]Until you set up your website to be deployed, LiveReload only updates the *local* version of your website. This version is only visible to you. In order to make your website searchable, discoverable, and live on the internet you will need to upload your website's files to a site builder. See Chapter 3 for details.

[4]You can also think of the RStudio Viewer as a web browser.

[5]Hugo has its own LiveReload implementation. If you want to take advantage of it, you may set the global option options(blogdown.generator.server = TRUE). See Section D.2 for more information.

[6]You can find its source on GitHub: https://github.com/yihui/hugo-lithium-theme. This theme was forked from https://github.com/jrutheiser/hugo-lithium-theme and modified to work better with **blogdown**.

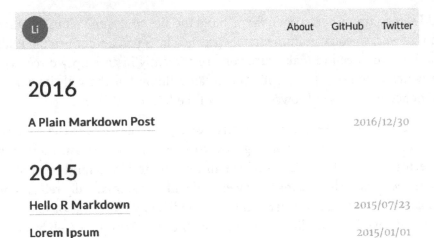

FIGURE 1.1: The homepage of the default new site.

able to change some obvious settings. For example, you may see configurations like these in `config.toml`:

```toml
baseurl = "/"
languageCode = "en-us"
title = "A Hugo website"
theme = "hugo-lithium-theme"

[[menu.main]]
    name = "About"
    url = "/about/"
[[menu.main]]
    name = "GitHub"
    url = "https://github.com/rstudio/blogdown"
[[menu.main]]
    name = "Twitter"
    url = "https://twitter.com/rstudio"
```

You can change the website title, e.g., `title = "My own cool website"`, and update the GitHub and Twitter URLs.

2. The content directory (by default, `content/`). This is where you write the R Markdown or Markdown source files for your posts and pages. Under `content/` of the default site, you can see `about.md` and a `post/` directory containing a few posts. The organization of the content directory is up to you. You can have arbitrary files and directories there, depending on the website structure you want.

3. The publishing directory (by default, `public/`). Your website will be generated to this directory, meaning that you do not need to manually add any files to this directory.[7] Typically it contains a lot of `*.html` files and dependencies like `*.css`, `*.js`, and images. You can upload everything under `public/` to any web server that can serve static websites, and your website will be up and running. There are many options for publishing static websites, and we will talk more about them in Chapter 3 if you are not familiar with deploying websites.

If you are satisfied with this default theme, you are basically ready to start writing and publishing your new website! We will show how to use other themes in Section 1.6. However, please keep in mind that a more complicated and fancier theme may require you to learn more about all the underlying technologies like the Hugo templating language, HTML, CSS, and JavaScript.

1.3 RStudio IDE

There are a few essential RStudio addins to make it easy to edit and preview your website, and you can find them in the menu "Addins" on the RStudio toolbar:

---

[7]By running either `serve_site()` or `build_site()`, files will be generated and published in your publishing directory automatically.

- "Serve Site": This addin calls `blogdown::serve_site()` to continuously serve your website locally using the LiveReload technology, so you can live preview the website. You can continue to edit material for your site while you are previewing it, but this function will block your R console by default, meaning that you will not be able to use your R console once you start this local web server. To unblock your console, click on the red stop sign in the top right corner of the console window. If you would rather avoid this behavior altogether, set the option `options(servr.daemon = TRUE)` before you click this addin or call the function `serve_site()`, so that the server is daemonized and will not block your R console.[8]

- "New Post": This addin provides a dialog box for you to enter the metadata of your blog post, including the title, author, date, and so on. See Figure 1.2 for an example. This addin actually calls the function `blogdown::new_post()` under the hood, but does a few things automatically:

  - As you type the title of the post, it will generate a filename for you, and you can edit it if you do not like the automatically generated one. In fact, you can also use this addin to create normal pages under any directories under `content/`. For example, if you want to add a resume page, you can change the filename to `resume.md` from the default `post/YYYY-mm-dd-resume.md`.

  - You can select the date from a calendar widget provided by Shiny.[9]

  - It will scan the categories and tags of existing posts, so when you want to input categories or tags, you can select them from the drop-down menus, or create new ones.

---

[8]We have heard of cases where the daemonized server crashed R on Windows. If you run into problems with the daemonized server, there are three workarounds, and you can try one of them: (1) install the **later** package via `install.packages("later")` and start the server again; (2) use Hugo's server (see Section D.2); (3) call `blogdown::serve_site()` in a separate R session, and you can preview your website in your web browser but can still edit the website in RStudio.

[9]Shiny is an R package for building interactive web apps using R. Using this addin, the calendar widget allows you to view an interactive calendar by month to select dates. This is a simple use of Shiny, but you can read more about Shiny apps here: `https://shiny.rstudio.com`.

– After a new post is created, it will be automatically opened, so you can start writing the content immediately.

- "Update Metadata": This addin allows you to update the YAML metadata of the currently opened post. See Figure 1.3 for an example. The main advantage of this addin is that you can select categories and tags from dropdown menus instead of having to remember them.

FIGURE 1.2: Create a new post using the RStudio addin.

With these addins, you should rarely need to run any R commands manually after you have set up your website, since all your posts will be automatically compiled whenever you create a new post or modify an existing post due to the LiveReload feature.

If your RStudio version is at least v1.1.383,[10] you can actually create a website project directly from the menu File -> New Project -> New Directory (see Figure 1.4 and 1.5).

If your website was created using the function blogdown::new_site() in-

---

[10]You may download all RStudio official releases including v1.1.383 from https://www. rstudio.com/products/rstudio/download/.

FIGURE 1.3: Update the metadata of an existing post using the RStudio addin.

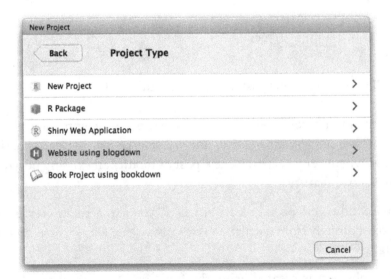

FIGURE 1.4: Create a new website project in RStudio.

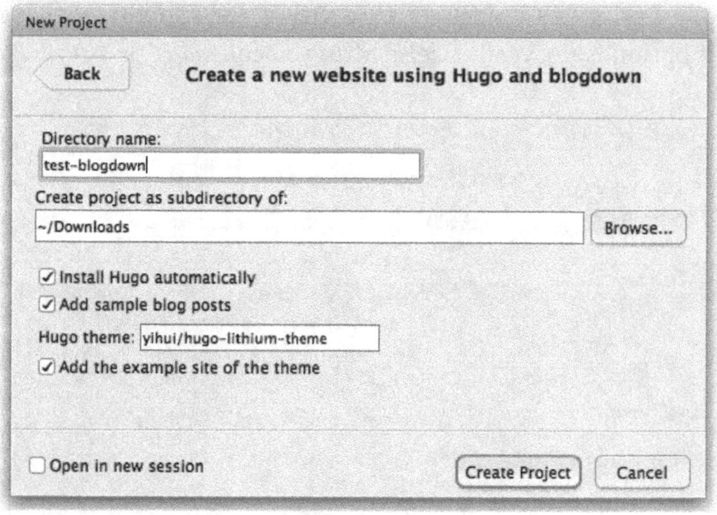

FIGURE 1.5: Create a website project based on blogdown.

stead of the RStudio menu for the first time, you can quit RStudio and open the project again. If you go to the menu Tools -> Project Options, your project type should be "Website" like what you can see in Figure 1.6.

Then you will see a pane in RStudio named "Build," and there is a button "Build Website." When you click this button, RStudio will call blogdown::build_site() to build the website. This will automatically generate files in the public/ directory.[11] If you want to build the website and publish the output files under the public/ manually, you are recommended to restart your R session and click this "Build Website" button every time before you publish the website, instead of publishing the public/ folder generated continuously and automatically by blogdown::serve_site(), because the latter calls blogdown::build_site(local = TRUE), which has some subtle differences with blogdown::build_site(local = FALSE) (see Section D.3 for details).

We strongly recommend that you uncheck the option "Preview site after building" in your RStudio project options (Figure 1.6).[12] You can also

---

[11]Or wherever your publishing directory is located. It is public/ by default, but it can be changed by specifying the publishDir = "myNewDirectory" in the config.toml file.

[12]In case you wonder why: unless you have set the option relativeurls to true in config.toml, it requires a web server to preview the website locally, otherwise even if you

uncheck the option "Re-knit current preview when supporting files change," since this option is not really useful after you call `serve_site()`.

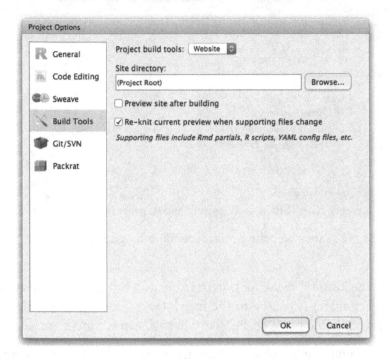

FIGURE 1.6: RStudio project options.

1.4 Global options

Depending on your personal preferences, you can set a few global options before you work on your website. These options should be set using `options(name = value)`, and currently available options are presented in Table 1.1.

We recommend that you set these options in your R startup profile file. You

can see the homepage of your website in the RStudio Viewer, most links like those links to CSS and JavaScript files are unlikely to work. When the RStudio Viewer shows you the preview, it does not actually launch a web server.

TABLE 1.1: Global options that affect the behavior of blogdown.

Option name	Default	Meaning
servr.daemon	FALSE	Whether to use a daemonized server
blogdown.author		The default author of new posts
blogdown.ext	.md	Default extension of new posts
blogdown.subdir	post	A subdirectory under content/
blogdown.yaml.empty	TRUE	Preserve empty fields in YAML?

can check out the help page `?Rprofile` for more details, and here is a simplified introduction. A startup profile file is basically an R script that is executed when your R session is started. This is a perfect place to set global options, so you do not need to type these options again every time you start a new R session. You can use a global profile file `~/.Rprofile`,[13] or a per-project file `.Rprofile` under the root directory of your RStudio project. The former will be applied to all R sessions that you start, unless you have provided the latter to override it. The easiest way to create such a file is to use `file.edit()` in RStudio, e.g.,

```
file.edit("~/.Rprofile")
# or file.edit('.Rprofile')
```

Suppose you always prefer the daemonized server and want the author of new posts to be "John Doe" by default. You can set these options in the profile file:

```
options(servr.daemon = TRUE, blogdown.author = "John Doe")
```

A nice consequence of setting these options is that when you use the RStudio addin "New Post," the fields "Author," "Subdirectory," and "Format" will be automatically populated, so you do not need to manipulate them every time unless you want to change the defaults (occasionally).

R only reads one startup profile file. For example, if you have a `.Rprofile` under the current directory and a global `~/.Rprofile`, only the former one will

---

[13]The tilde ~ denotes your home directory in your system.

be executed when R starts up from the current directory. This may make it inconvenient for multiple authors collaborating on the same website project, since you cannot set author-specific options. In particular, it is not possible to set the `blogdown.author` option in a single `.Rprofile`, because this option should be different for different authors. One workaround is to set common options in `.Rprofile` under the root directory of the website project, and also execute the global `~/.Rprofile` if it exists. Author-specific options can be set in the global `~/.Rprofile` on each author's computer.

```r
# in .Rprofile of the website project
if (file.exists("~/.Rprofile")) {
  base::sys.source("~/.Rprofile", envir = environment())
}
# then set options(blogdown.author = 'Your Name') in
# ~/.Rprofile
```

1.5   R Markdown vs. Markdown

If you are not familiar with R Markdown, please see Appendix A for a quick tutorial. When you create a new post, you have to decide whether you want to use R Markdown or plain Markdown, as you can see from Figure 1.2. The main differences are:

1. You cannot execute any R code in a plain Markdown document, whereas in an R Markdown document, you can embed R code chunks (```` ```{r} ````). However, you can still embed R code in plain Markdown using the syntax for fenced code blocks ```` ```r ```` (note there are no curly braces `{}`). Such code blocks will not be executed and may be suitable for pure demonstration purposes. Below is an example of an R code chunk in R Markdown:

```
```{r cool-plot, fig.width='80%', fig.cap='A cool plot.'}
plot(cars, pch = 20)  # not really cool
```
```

And here is an example of an R code block in plain Markdown:

```
```r

1 + 1  # not executed
```
```

2. A plain Markdown post is rendered to HTML through Blackfriday[14] (a package written in the Go language and adopted by Hugo). An R Markdown document is compiled through the packages **rmarkdown**, **bookdown**, and Pandoc, which means you can use most features of Pandoc's Markdown[15] and **bookdown**'s Markdown extensions[16] in **blogdown**. If you use R Markdown (Allaire et al., 2017) with **blogdown**, we recommend that you read the documentation of Pandoc and **bookdown** at least once to know all the possible features. We will not repeat the details in this book, but list the features briefly below, which are also demonstrated on the example website `https://blogdown-demo.rbind.io`.

    - Inline formatting: `_italic_` / `**bold**` text and `` `inline code` ``.

    - Inline elements: subscripts (e.g., `H~2~0`) and superscripts (e.g., `R^2^`); links (`[text](url)`) and images `![title](url)`; footnotes `text^[footnote]`.

    - Block-level elements: paragraphs; numbered and unnumbered section headers; ordered and unordered lists; block quotations; fenced code blocks; tables; horizontal rules.

    - Math expressions and equations.

---

[14]`https://gohugo.io/overview/configuration/`
[15]`http://pandoc.org/MANUAL.html#pandocs-markdown`
[16]`https://bookdown.org/yihui/bookdown/components.html`

- Theorems and proofs.

- R code blocks that can be used to produce text output (including tables) and graphics. Note that equations, theorems, tables, and figures can be numbered and cross-referenced.

- Citations and bibliography.

- HTML widgets, and Shiny apps embedded via `<iframe>`.

There are many differences in syntax between Blackfriday's Markdown and Pandoc's Markdown. For example, you can write a task list with Blackfriday but not with Pandoc:

```
- [x] Write an R package.
- [ ] Write a book.
- [ ] ...
- [ ] Profit!
```

Similarly, Blackfriday does not support LaTeX math and Pandoc does. We have added the MathJax[17] support to the default theme (hugo-lithium-theme[18]) in **blogdown** to render LaTeX math on HTML pages, but there is a caveat for plain Markdown posts: you have to include inline math expressions in a pair of backticks `` `$math$` ``, e.g., `` `$S_n = \sum_{i=1}^n X_i$` ``. Similarly, math expressions of the display style have to be written in `` `$$math$$` ``. For R Markdown posts, you can use `$math$` for inline math expressions, and `$$math$$` for display-style expressions.[19]

If you find it is a pain to have to remember the differences between R Markdown and Markdown, a conservative choice is to always use R Markdown, even if your document does not contain any R code chunks. Pandoc's Mark-

---

[17] https://www.mathjax.org/#docs

[18] https://github.com/yihui/hugo-lithium-theme

[19] The reason that we need the backticks for plain Markdown documents is that we have to prevent the LaTeX code from being interpreted as Markdown by Blackfriday. Backticks will make sure the inner content is not translated as Markdown to HTML, e.g., `` `$$x *y* z$$` `` will be converted to `<code>$$x *y* z$$</code>`. Without the backticks, it will be converted to `$$x <em>y</em> z$$`, which is not a valid LaTeX math expression for MathJax. Similar issues can arise when you have other special characters like underscores in your math expressions.

down is much richer than Blackfriday, and there are only a small number of features unavailable in Pandoc but present in Blackfriday. The main disadvantages of using R Markdown are:

1. You may sacrifice some speed in rendering the website, but this may not be noticeable due to a caching mechanism in **blogdown** (more on this in Section D.3). Hugo is very fast when processing plain Markdown files, and typically it should take less than one second to render a few hundred Markdown files.

2. You will have some intermediate HTML files in the source directory of your website, because **blogdown** has to call **rmarkdown** to pre-render *.Rmd files *.html. You will also have intermediate folders for figures (*_files/) and cache (*_cache/) if you have plot output in R code chunks or have enabled **knitr**'s caching. Unless you care a lot about the "cleanness" of the source repository of your website (especially when you use a version control tool like GIT), these intermediate files should not matter.

In this book, we usually mean .Rmd files when we say "R Markdown documents," which are compiled to .html by default. However, there is another type of R Markdown document with the filename extension .Rmarkdown. Such R Markdown documents are compiled to Markdown documents with the extension .markdown, which will be processed by Hugo instead of Pandoc. There are two major limitations of using .Rmarkdown compared to .Rmd:

- You cannot use Markdown features only supported by Pandoc, such as citations. Math expressions only work if you have installed the **xaringan** package (Xie, 2017e) and applied the JavaScript solution mentioned in Section B.3.

- HTML widgets are not supported.

The main advantage of using .Rmarkdown is that the output files are cleaner because they are Markdown files. It can be easier for you to read the output of your posts without looking at the actual web pages rendered. This can be particularly helpful when reviewing GitHub pull requests. Note that numbered tables, figures, equations, and theorems are also supported. You cannot directly use Markdown syntax in table or figure captions, but you can use text references as a workaround (see **bookdown**'s documentation).

For any R Markdown documents (not specific to **blogdown**), you have to specify an output format. There are many possible output formats[20] in the **rmarkdown** package (such as `html_document` and `pdf_document`) and other extension packages (such as `tufte::tufte_html` and `bookdown::gitbook`). Of course, the output format for websites should be HTML. We have provided an output format function `blogdown::html_page` in **blogdown**, and all R Markdown files are rendered using this format. It is based on the output format `bookdown::html_document2`, which means it has inherited a lot of features from **bookdown** in addition to features in Pandoc. For example, you can number and cross-reference math equations, figures, tables, and theorems, etc. See Chapter 2 of the **bookdown** book (Xie, 2016) for more details on the syntax.

Note that the output format `bookdown::html_document2` in turn inherits from `rmarkdown::html_document`, so you need to see the help page `?rmarkdown::html_document` for all possible options for the format `blogdown::html_page`. If you want to change the default values of the options of this output format, you can add an `output` field to your YAML metadata. For example, we can add a table of contents to a page, set the figure width to be 6 inches, and use the `svg` device for plots by setting these options in YAML:

```
---
title: "My Awesome Post"
author: "John Doe"
date: "2017-02-14"
output:
  blogdown::html_page:
    toc: true
    fig_width: 6
    dev: "svg"
---
```

To set options for `blogdown::html_page()` globally (i.e., apply certain options

---

[20] http://rmarkdown.rstudio.com/lesson-9.html

to all Rmd files), you can create a `_output.yml` file under the root directory of your website. This YAML file should contain the output format directly (do not put the output format under the `output` option), e.g.,

```
blogdown::html_page:
  toc: true
  fig_width: 6
  dev: "svg"
```

At the moment, not all features of `rmarkdown::html_document` are supported in **blogdown**, such as `df_print`, `code_folding`, `code_download`, and so on.

If your code chunk has graphics output, we recommend that you avoid special characters like spaces in the chunk label. Ideally, you should only use alphanumeric characters and dashes, e.g., ```` ```{r, my-label} ```` instead of ```` ```{r, my label} ````.

It is not recommended to change the **knitr** chunk options `fig.path` or `cache.path` in R Markdown. The default values of these options work best with **blogdown**. Please read Section D.5 to know the technical reasons if you prefer.

If you are working on an R Markdown post, but do not want **blogdown** to compile it, you can temporarily change its filename extension from `.Rmd` to another unknown extension such as `.Rmkd`.

1.6 Other themes

In Hugo, themes control the entire appearance and functionality of your site. So, if you care a lot about the appearance of your website, you will probably spend quite a bit of time in the beginning looking for a Hugo theme that you like from the collection listed at `http://themes.gohugo.io`. Please note that not all themes have been tested against **blogdown**. If you find a certain theme does not work well with **blogdown**, you may report to `https://github.com/rstudio/blogdown/issues`, and we will try to investigate the reason, but it can be time-consuming to learn and understand how a new theme works, so we recommend that you learn more about Hugo by yourself before asking, and we also encourage users to help each other there.

After you have found a satisfactory theme, you need to figure out its GitHub username and repository name,[21] then either install the theme via `blogdown::install_theme()`, or just create a new site under another new directory and pass the GitHub repository name to the `theme` argument of `new_site()`. We recommend that you use the second approach, because Hugo themes could be very complicated and the usage of each theme can be very different and highly dependent on `config.toml`. If you install a theme using `install_theme()` instead of `new_site()` you'll need to manually create the `config.toml` file in the root directory of your website to match the newly installed theme.[22]

```
# for example, create a new site with the academic theme
blogdown::new_site(theme = "gcushen/hugo-academic")
```

To save you some time, we list a few themes below that match our taste:

- Simple/minimal themes: XMin,[23] simple-a,[24] and ghostwriter.[25]

- Sophisticated themes: hugo-academic[26] (strongly recommended for users in academia), hugo-tranquilpeak-theme,[27] hugo-creative-portfolio-theme,[28] and hugo-universal-theme.[29]

- Multimedia content themes: If you are interested in adding multimedia

---

[21] For most themes, you can find this by navigating to the theme of your choice from `http://themes.gohugo.io` and then clicking on `Homepage`.

[22] In a workaround, if you used `install_theme()` and set the `theme_example` argument to TRUE, then you can access an example `config.toml` file. In the `themes/` directory, navigate to the file for your newly downloaded theme and find `exampleSite/config.toml`. This file can be copied to your root directory (to replace the `config.toml` file from your original theme) or used as a template to correctly write a new `config.toml` file for your new theme.

[23] `https://github.com/yihui/hugo-xmin`
[24] `https://github.com/AlexFinn/simple-a`
[25] `https://github.com/jbub/ghostwriter`
[26] `https://github.com/gcushen/hugo-academic`
[27] `https://github.com/kakawait/hugo-tranquilpeak-theme`
[28] `https://github.com/kishaningithub/hugo-creative-portfolio-theme`
[29] `https://github.com/devcows/hugo-universal-theme`

content to your site (such as audio files of a podcast), the castanet[30] theme provides an excellent framework tailored for this application. An example of a site using **blogdown** with the castanet theme is the R-Podcast.[31]

If you do not understand HTML, CSS, or JavaScript, and have no experience with Hugo themes or templates, it may take you about 10 minutes to get started with your new website, since you have to accept everything you are given (such as the default theme); if you do have the knowledge and experience (and desire to highly customize your site), it may take you several days to get started. Hugo is really powerful. Be cautious with power.

Another thing to keep in mind is that the more effort you make in a complicated theme, the more difficult it is to switch to other themes in the future, because you may have customized a lot of things that are not straightforward to port to another theme. So please ask yourself seriously, "Do I like this fancy theme so much that I will definitely not change it in the next couple of years?"

If you choose to dig a rather deep hole, someday you will have no choice but keep on digging, even with tears.

— Liyun Chen[32]

1.7 A recommended workflow

There are a lot of ways to start building a website and deploy it. Because of the sheer number of technologies that you need to learn to fully understand how a website works, we'd like to recommend one workflow to beginners,

---

[30]https://github.com/mattstratton/castanet
[31]https://www.r-podcast.org
[32]Translated from her Chinese Weibo: http://weibo.com/1406511850/Dhrb4toHc (you cannot view this page unless you have logged in).

so that hopefully they do not need to digest the rest of this book. This is definitely not the most optimal workflow, but requires you to know the fewest technical details.

To start a new website:

1. Carefully pick a theme at `http://themes.gohugo.io`, and find the link to its GitHub repository, which is of the form `https://github.com/user/repo`.

2. Create a new project in RStudio, and type the code `blogdown::new_site(theme = 'user/repo')` in the R console, where `user/repo` is from the link in Step 1.

3. Play with the new site for a while and if you do not like it, you can repeat the above steps, otherwise edit the options in `config.toml`. If you do not understand certain options, go to the documentation of the theme, which is often the README page of the GitHub repository. Not all options have to be changed.

To edit a website:

1. Set `options(servr.daemon = TRUE)` unless you have already set it in `.Rprofile`. If this option does not work for you (e.g., it crashes your R session), see Section 1.4 for a workaround.

2. Click the RStudio addin "Serve Site" to preview the site in RStudio Viewer. This only needs to be done once every time you open the RStudio project or restart your R session. Do not click the Knit button on the RStudio toolbar.

3. Use the "New Post" addin to create a new post or page, then start writing the content.

4. Use the "Update Metadata" addin to modify the YAML metadata if necessary.

To publish a website:

1. Restart the R session, and run `blogdown::hugo_build()`. You

should get a `public/` directory under the root directory of your project.

2. Log into `https://www.netlify.com` (you can use a GitHub account if you have one). If this is the first time you have published this website, you can create a new site, otherwise you may update the existing site you created last time. You can drag and drop the `public/` folder from your file viewer to the indicated area on the Netlify web page, where it says "Drag a folder with a static site here."

3. Wait for a few seconds for Netlify to deploy the files, and it will assign a random subdomain of the form `random-word-12345.netlify.com` to you. You can (and should) change this random subdomain to a more meaningful one if it is still available.

If you are familiar with GIT and GitHub, we recommend that you create a new site from your GitHub repository that contains the source files of your website, so that you can enjoy the benefits of continuous deployment instead of manually uploading the folder every time. See Chapter 3 for more information.

2

Hugo

In this chapter, we will briefly introduce Hugo (`https://gohugo.io`), the static site generator on which **blogdown** is based. This chapter is not meant to replace the official Hugo documentation, but provide a guide to those who are just getting started with Hugo. When in doubt, please consult the official Hugo documentation.

2.1 Static sites and Hugo

A static site often consists of HTML files (with optional external dependencies like images and JavaScript libraries), and the web server sends exactly the same content to the web browser no matter who visits the web pages. There is no dynamic computing on the server when a page is requested. By comparison, a dynamic site relies on a server-side language to do certain computing and sends potentially different content depending on different conditions. A common language is PHP, and a typical example of a dynamic site is a web forum. For example, each user has a profile page, but typically this does not mean the server has stored a different HTML profile page for every single user. Instead, the server will fetch the user data from a database, and render the profile page dynamically.

For a static site, each URL you visit often has a corresponding HTML file stored on the server, so there is no need to compute anything before serving the file to visitors. This means static sites tend to be faster in response time than dynamic sites, and they are also much easier to deploy, since deployment simply means copying static files to a server. A dynamic site often relies on databases, and you will have to install more software packages to serve a dynamic site. For more advantages of static sites, please read the introduction[1] on Hugo's website.

There are many existing static site generators, including Hugo, Jekyll,[2] and Hexo,[3] etc. Most of them can build general-purpose websites but are often used to build blogs.

We love Hugo for many reasons, but there are a few that stand out. Unlike other static site generators, the installation of Hugo is very simple because it provides a single executable without dependencies for most operating systems (see Section 1.1). It was also designed to render hundreds of pages of content faster than comparable static site generators and can reportedly render a single page in approximately 1 millisecond. Lastly, the community of Hugo users is very active both on the Hugo discussion forum[4] and on GitHub issues.[5]

Although we think Hugo is a fantastic static site generator, there is really one and only one major missing feature: the support for R Markdown. That is basically the whole point of the **blogdown** package.[6] This missing feature means that you cannot easily generate results using R code on your web pages, since you can only use static Markdown documents. Besides, you have to use Hugo's choice of the Markdown engine named "Blackfriday" instead of the more powerful Pandoc.[7]

Hugo uses a special file and folder structure to create your website (Figure

---

[1] https://gohugo.io/overview/introduction/

[2] http://jekyllrb.com

[3] https://hexo.io

[4] https://discuss.gohugo.io

[5] https://github.com/gohugoio/hugo/issues

[6] Another motivation was an easier way to create new pages or posts. Static site generators often provide commands to create new posts, but you often have to open and modify the new file created by hand after using these commands. I was very frustrated by this, because I was looking for a graphical user interface where I can just fill out the title, author, date, and other information about a page, then I can start writing the content right away. That is why I provided the RStudio addin "New Post" and the function `blogdown::new_post()`. In the past few years, I hated it every time I was about to create a new post either by hand or via the Jekyll command line. Finally, I felt addicted to blogging again after I finished the RStudio addin.

[7] There is a feature request in Hugo's repository that is more than three years old: `https://github.com/gohugoio/hugo/issues/234`, and it seems that it will not be implemented in the near future.

2.1). The rest of this chapter will give more details on the following files and folders:

- `config.toml`
- `content/`
- `static/`
- `themes/`
- `layouts/`

FIGURE 2.1: Possible files and folders created when you create a new site using **blogdown**.

2.2 Configuration

The first file that you may want to look at is the configuration or `config` file in your root directory, in which you can set global configurations of your site. It may contain options like the title and description of your site, as well as other global options like links to your social networks, the navigation menu, and the base URL for your website.

When generating your site, Hugo will search for a file called `config.toml` first. If it cannot find one, it will continue to search for `config.yaml`.[8] Since

---

[8]Hugo also supports `config.json`, but **blogdown** does not support it, so we do not recommend that you use it.

most Hugo themes contain example sites that ship `config.toml` files, and
the TOML (Tom's Obvious, Minimal Language) format appears to be more
popular in the Hugo community, we will mainly discuss `config.toml` here.

We recommend that you use the TOML syntax only for the config file (you
can also use YAML if you prefer), and use YAML as the data format for
the metadata of (R) Markdown pages and posts, because R Markdown and
blogdown fully support only YAML.[9] If you have a website that has already
used TOML, you may use `blogdown::hugo_convert(unsafe = TRUE)` to convert TOML data to YAML, but please first make sure you have backed up the
website because it will overwrite your Markdown files.

The Hugo documentation does not use TOML or YAML consistently in its
examples, which can be confusing. Please pay close attention to the configuration format when copying examples to your own website.

2.2.1 TOML Syntax

If you are not familiar with the TOML Syntax, we will give a brief overview
and you may read the full documentation[10] to know the details.

TOML is made up of key-value pairs separated by equal signs:

```
key = value
```

When you want to edit a configuration in the TOML file, simply change the
value. Values that are character strings should be in quotes, whereas Boolean
values should be lowercase and bare.

For example, if you want to give your website the title "My Awesome Site,"
and use relative URLs instead of the default absolute URLs, you may have
the following entries in your `config.toml` file.

---

[9]TOML has its advantages, but I feel they are not significant in the context of Hugo
websites. It is a pain to have to know yet another language, TOML, when YAML stands for
"Yet Another Markup Language." I'm not sure if the XKCD comic applies in this case: `https:
//xkcd.com/927/`.

[10]`https://github.com/toml-lang/toml`

```
title = "My Awesome Site"

relativeURLs = true
```

Most of your website's global variables are entered in the `config.toml` file in exactly this manner.

Further into your `config` file, you may notice some values in brackets like this:

```
[social]
    github  = "https://github.com/rstudio/blogdown"
    twitter = "https://twitter.com/rstudio"
```

This is a table in the TOML language and Hugo uses them to fill in information on other pages within your site. For instance, the above table will populate the `.Site.Social` variable in your site's templates (more information on this in Section 2.5).

Lastly, you may find some values in double brackets like this:

```
[[menu.main]]
    name = "Blog"
    url = "/blog/"

[[menu.main]]
    name = "Categories"
    url = "/categories/"

[[menu.main]]
    name = "About"
    url = "/about/"
```

In TOML, double brackets are used to indicate an array of tables. Hugo interprets this information as a menu. If the code above was found in a `config.toml` file, the resulting website would have links to Blog, Categories, and About pages in the site's main menu. The location and styling of that menu are specified elsewhere, but the names of each menu's choices and the links to each section are defined here.

The `config.toml` file is different for each theme. Make sure that when you choose a theme, you read its documentation thoroughly to get an understanding of what each of the configuration options does (more on themes in Section 2.4).

2.2.2 Options

All built-in options that you may set for Hugo are listed at `https://gohugo.io/overview/configuration/`. You can change any of these options except `contentDir`, which is hard-coded to `content` in **blogdown**. Our general recommendation is that you'd better not modify the defaults unless you understand the consequences. We list a few options that may be of interest to you:

- `baseURL`: Normally you have to change the value of this option to the base URL of your website. Some Hugo themes may have it set to `http://replace-this-with-your-hugo-site.com/` or `http://www.example.com/` in their example sites, but please make sure to replace them with your own URL (see Chapter 3 and Appendix C for more information on publishing websites and obtaining domain names). Note that this option can be a URL with a subpath, if your website is to be published under a subpath of a domain name, e.g., `http://www.example.com/docs/`.

- `enableEmoji`: You may set it to `true` so that you can use Emoji emoticons[11] like `:smile:` in Markdown.

- `permalinks`: Rules to generate permanent links of your pages. By default, Hugo uses full filenames under `content/` to generate links, e.g., `content/about.md` will be rendered to `public/about/index.html`, and `content/post/2015-07-23-foo.md` will be rendered to `public/post/2015-07-23-foo/index.html`, so the actual links are `/about/` and `/post/2015-07-23-foo/` on the website. Although it is not required to set custom rules for permanent links, it is common to see links of the form `/YYYY/mm/dd/post-title/`. Hugo allows you to use several pieces of

---

[11] `http://www.emoji-cheat-sheet.com`

information about a source file to generate a link, such as the date (year, month, and day), title, and filename, etc. The link can be independent of the actual filename. For example, you may ask Hugo to render pages under content/post/ using the date and title for their links:

```
[permalinks]
    post = "/:year/:month/:day/:title/"
```

Personally, I recommend that you use the :slug variable[12] instead of :title:

```
[permalinks]
    post = "/:year/:month/:day/:slug/"
```

This is because your post title may change, and you probably do not want the link to the post to change, otherwise you have to redirect the old link to the new link, and there will other types of trouble like Disqus comments. The :slug variable falls back to :title if a field named slug is not set in the YAML metadata of the post. You can set a fixed slug so that the link to the post is always fixed and you will have the freedom to update the title of your post.

You may find a list of all possible variables that you can use in the permalinks option at https://gohugo.io/extras/permalinks/.

- publishDir: The directory under which you want to generate the website.

- theme: The directory name of the Hugo theme under themes/.

- ignoreFiles: A list of filename patterns (regular expressions) for Hugo to ignore certain files when building the site. I recommend that you specify at least these patterns ["\\.Rmd$", "\\.Rmarkdown$", "_files$", "_cache$"]. You should ignore .Rmd files because **blogdown** will compile

---

[12]A slug is simply a character string that you can use to identify a specific post. A slug will not change, even if the title changes. For instance, if you decide to change the title of your post from "I love blogdown" to "Why blogdown is the best package ever," and you used the post's title in the URL, your old links will now be broken. If instead, you specified the URL via a slug (something like "blogdown-love"), then you can change the title as many times as you'd like and you will not end up with any broken links.

them to .html, and it suffices for Hugo to use the .html files. There is no
need for Hugo to build .Rmd files, and actually Hugo does not know how. Directories with suffixes _files and _cache should be ignored because they
contain auxiliary files after an Rmd file is compiled, and **blogdown** will
store them. Hugo should not copy them again to the public/ directory.

- uglyURLs: By default, Hugo generates "clean" URLs. This may be a little surprising and requires that you understand how URLs work when
  your browser fetches a page from a server. Basically, Hugo generates
  foo/index.html for foo.md by default instead of foo.html, because the former allows you to visit the page via the clean URL foo/ without index.html.
  Most web servers understand requests like http://www.example.com/foo/
  and will present index.html under foo/ to you. If you prefer the strict map-
  ping from *.md to *.html, you may enable "ugly" URLs by setting uglyURLs
  to true.

- hasCJKLanguage: If your website is primarily in CJK (Chinese, Korean, and
  Japanese), I recommend that you set this option to true, so that Hugo's
  automatic summary and word count work better.

Besides the built-in Hugo options, you can set other arbitrary options in
config.toml. For example, it is very common to see an option named params,
which is widely used in many Hugo themes. When you see a variable
.Site.Params.FOO in a Hugo theme, it means an option FOO that you set under [params] in config.toml, e.g., .Site.Params.author is Frida Gomam with
the following config file:

```
[params]
    author = "Frida Gomam"
    dateFormat = "2006/01/02"
```

The goal of all these options is to avoid hard-coding anything in Hugo
themes, so that users can easily edit a single config file to apply the theme
to their websites, instead of going through many HTML files and making
changes one by one.

2.3 Content

The structure of the `content/` directory can be arbitrary. A common structure is that there are a few static pages under the root of `content/`, and a subdirectory `post/` containing blog posts:

```
├── _index.md
├── about.md
├── vitae.md
├── post/
│   ├── 2017-01-01-foo.md
│   ├── 2017-01-02-bar.md
│   └── ...
└── ...
```

2.3.1 YAML metadata

Each page should start with YAML metadata specifying information like the title, date, author, categories, tags, and so on. Depending on the specific Hugo theme and templates you use, some of these fields may be optional.

Among all YAML fields, we want to bring these to your attention:

- `draft`: You can mark a document as a draft by setting `draft: true` in its YAML metadata. Draft posts will not be rendered if the site is built via `blogdown::build_site()` or `blogdown::hugo_build()`, but will be rendered in the local preview mode (see Section D.3).

- `publishdate`: You may specify a future date to publish a post. Similar to draft posts, future posts are only rendered in the local preview mode.

- `weight`: This field can take a numeric value to tell Hugo the order of pages when sorting them, e.g., when you generate a list of all pages under a directory, and two posts have the same date, you may assign different weights to them to get your desired order on the list.

- slug: A character string as the tail of the URL. It is particularly useful when you define custom rules for permanent URLs (see Section 2.2.2).

2.3.2 Body

As we mentioned in Section 1.5, your post can be written in either R Markdown or plain Markdown. Please be cautious about the syntax differences between the two formats when you write the body of a post.

2.3.3 Shortcode

Besides all Markdown features, Hugo provides a useful feature named "shortcodes." You can use a shortcode in the body of your post. When Hugo renders the post, it can automatically generate an HTML snippet based on the parameters you pass to the shortcode. This is convenient because you do not have to type or embed a large amount of HTML code in your post. For example, Hugo has a built-in shortcode for embedding Twitter cards. Normally, this is how you embed a Twitter card (Figure 2.2) on a page:

```
<blockquote class="twitter-tweet">
  <p lang="en" dir="ltr">Anyone know of an R package for
  interfacing with Alexa Skills?
    <a href="https://twitter.com/thosjleeper">@thosjleeper</a>
    <a href="https://twitter.com/xieyihui">@xieyihui</a>
    <a href="https://twitter.com/drob">@drob</a>
    <a href="https://twitter.com/JennyBryan">@JennyBryan</a>
    <a href="https://twitter.com/HoloMarkeD">@HoloMarkeD</a> ?
  </p>
  &mdash; Jeff Leek (@jtleek)
  <a href="https://twitter.com/jtleek/status/852205086956818432">
    April 12, 2017
  </a>
</blockquote>
<script async src="//platform.twitter.com/widgets.js" charset="utf-8">
</script>
```

FIGURE 2.2: A tweet by Jeff Leek.

If you use the shortcode, all you need in the Markdown source document is:

```
{{< tweet 852205086956818432 >}}
```

Basically, you only need to pass the ID of the tweet to a shortcode named tweet. Hugo will fetch the tweet automatically and render the HTML snippet for you. For more about shortcodes, see https://gohugo.io/extras/shortcodes/.

Shortcodes are supposed to work in plain Markdown documents only. To use shortcodes in R Markdown instead of plain Markdown, you have to call the function blogdown::shortcode(), e.g.,

```
```{r echo=FALSE}
blogdown::shortcode('tweet', '852205086956818432')
```
```

2.4   Themes

A Hugo theme is a collection of template files and optional website assets such as CSS and JavaScript files. In a nutshell, a theme defines what your website looks like after your source content is rendered through the templates.

Hugo has provided a large number of user-contributed themes at `https:` `//themes.gohugo.io`. Unless you are an experienced web designer, you'd better start from an existing theme here. The quality and complexity of these themes vary a lot, and you should choose one with caution. For example, you may take a look at the number of stars of a theme repository on GitHub, as well as whether the repository is still relatively active. We do not recommend that you use a theme that has not been updated for more than a year.

In this section, we will explain how the default theme in **blogdown** works, which may also give you some ideas about how to get started with other themes.

2.4.1 The default theme

The default theme in **blogdown**, hugo-lithium-theme, is hosted on GitHub at `https://github.com/yihui/hugo-lithium-theme`. It was originally written by Jonathan Rutheiser, and I have made several changes to it. This theme is suitable for those who prefer minimal styles, and want to build a website with a few pages and some blog posts.

Typically a theme repository on GitHub has a `README` file, which also serves as the documentation of the theme. After you read it, the next file to look for is `config.toml` under the `exampleSite` directory, which contains sample configurations for a website based on this theme. If a theme does not have a `README` file or `exampleSite` directory, you probably should not use it.

The `config.toml` of the theme hugo-lithium-theme contains the following options:

```
baseurl = "/"
relativeurls = false
languageCode = "en-us"
title = "A Hugo website"
theme = "hugo-lithium-theme"
googleAnalytics = ""
disqusShortname = ""
ignoreFiles = ["\\.Rmd$", "\\.Rmarkdown", "_files$", "_cache$"]
```

```
[permalinks]
    post = "/:year/:month/:day/:slug/"

[[menu.main]]
    name = "About"
    url = "/about/"
[[menu.main]]
    name = "GitHub"
    url = "https://github.com/rstudio/blogdown"
[[menu.main]]
    name = "Twitter"
    url = "https://twitter.com/rstudio"

[params]
    description = "A website built through Hugo and blogdown."

    highlightjsVersion = "9.11.0"
    highlightjsCDN = "//cdn.bootcss.com"
    highlightjsLang = ["r", "yaml"]
    highlightjsTheme = "github"

    MathJaxCDN = "//cdn.bootcss.com"
    MathJaxVersion = "2.7.1"

[params.logo]
    url = "logo.png"
    width = 50
    height = 50
    alt = "Logo"
```

Some of these options may be obvious to understand, and some may need explanations:

- baseurl: You can configure this option later, after you have a domain name for your website. Do not forget the trailing slash.

- relativeurls: This is optional. You may want to set it to true only if you intend to view your website locally through your file viewer, e.g., double-click on an HTML file and view it in your browser. This option defaults to `false` in Hugo, and it means your website must be viewed through a web server, e.g., `blogdown::serve_site()` has provided a local web server, so you can preview your website locally when `relativeurls = false`.

- `title`: The title of your website. Typically this is displayed in a web browser's title bar or on a page tab.

- `theme`: The directory name of the theme. You need to be very careful when changing themes, because one theme can be drastically different from another theme in terms of the configurations. It is quite possible that a different theme will not work with your current `config.toml`. Again, you have to read the documentation of a theme to know what options are supported or required.

- `googleAnalytics`: The Google Analytics tracking ID (like `UA-000000-2`). You can sign up at `https://analytics.google.com` to obtain a tracking ID.

- `disqusShortname`: The Disqus ID that you created during the account setup process at `https://disqus.com`. This is required to enable commenting on your site.[13] Please note that you have to set up a functional `baseurl` and publish your website before Disqus comments can work.

- `ignoreFiles` and `permalinks`: These options have been explained in Section 2.2.2.

- `menu`: This list of options specifies the text and URL of menu items at the top. See Figure 1.1 for a sample page. You can change or add more menu items. If you want to order the items, you may assign a `weight` to each item, e.g.,

```
[[menu.main]]
    name = "Home"
    url = "/"
    weight = 1
```

---

[13]As we mentioned in Section 2.1, **blogdown** generates static and unchanging content. To add something dynamic and always changing (like the ability for your fans to leave comments), you must incorporate an outside commenting system like Disqus.

```
[[menu.main]]
    name = "About"
    url = "/about/"
    weight = 2
[[menu.main]]
    name = "GitHub"
    url = "https://github.com/rstudio/blogdown"
    weight = 3
[[menu.main]]
    name = "CV"
    url = "/vitae/"
    weight = 4
[[menu.main]]
    name = "Twitter"
    url = "https://twitter.com/rstudio"
    weight = 5
```

In the above example, I added a menu item CV with the URL /vitae/, and
there is supposed to be a corresponding source file vitae.md under the
content/ directory to generate the page /vitae/index.html, so the link will
actually function.

- params: Miscellaneous parameters for the theme.

    - description: A brief description of your website. It is not visible on
      web pages (you can only see it from the HTML source), but should
      give search engines a hint about your website.

    - highlightjs*: These options are used to configure the JavaScript
      library highlight.js[14] for syntax highlighting of code blocks on the
      web pages. You can change the version (e.g., 9.12.0), the CND
      host (e.g., using cdnjs[15]: //cdnjs.cloudflare.com/ajax/libs), add
      more languages (e.g., ["r", "yaml", "tex"]), and change the theme
      (e.g., atom-one-light). See https://highlightjs.org/static/demo/
      for all languages and themes that highlight.js supports.

---

[14]https://highlightjs.org
[15]https://cdnjs.com

- `MathJax*`: The JavaScript library MathJax can render LaTeX math expressions on web pages. Similar to `highlightjsCDN`, you can specify the CDN host of MathJax, e.g., `//cdnjs.cloudflare.com/ajax/libs`, and you can also specify the version of MathJax.

- `logo`: A list of options to define the logo of the website. By default, the image `logo.png` under the `static/` directory is used.

If you want to be a theme developer and fully understand all the technical details about these options, you have to understand Hugo templates, which we will introduce in Section 2.5.

2.5   Templates

A Hugo theme consists of two major components: templates, and web assets. The former is essential, and tells Hugo how to render a page.[16] The latter is optional but also important. It typically consists of CSS and JavaScript files, as well as other assets like images and videos. These assets determine the appearance and functionality of your website, and some may be embedded in the content of your web pages.

You can learn more about Hugo templates from the official documentation (`https://gohugo.io/templates/overview/`). There are a great many different types of templates. To make it easier for you to master the key ideas, I created a very minimal Hugo theme, which covers most functionalities that an average user may need, but the total number of lines is only about 150, so we can talk about all the source code of this theme in the following subsection.

---

[16]The most common functionality of templates is to render HTML pages, but there can also be special templates, for example, for RSS feeds and sitemaps, which are XML files.

2.5.1 A minimal example

XMin[17] is a Hugo theme I wrote from scratch in about 12 hours. Roughly half an hour was spent on templates, 3.5 hours were spent on tweaking the CSS styles, and 8 hours were spent on the documentation (`https://xmin.yihui.name`). I think this may be a representative case of how much time you would spend on each part when designing a theme. It is perhaps our nature to spend much more time on cosmetic stuff like CSS than essential stuff like templates. Meanwhile, coding is often easier than documentation.

We will show the source code of the XMin theme. Because the theme may be updated occasionally in the future, you may follow this link to obtain a fixed version that we will talk about in this section: `https://github.com/yihui/hugo-xmin/tree/4bb305`. Below is a tree view of all files and directories in the theme:

```
hugo-xmin/
├── LICENSE.md
├── README.md
├── archetypes
│   └── default.md
├── layouts
│   ├── 404.html
│   ├── _default
│   │   ├── list.html
│   │   ├── single.html
│   │   └── terms.html
│   └── partials
│       ├── foot_custom.html
│       ├── footer.html
│       ├── head_custom.html
│       └── header.html
├── static
│   └── css
│       ├── fonts.css
│       └── style.css
```

---

[17]`https://github.com/yihui/hugo-xmin`

```
└── exampleSite
    ├── config.toml
    ├── content
    │   ├── _index.md
    │   ├── about.md
    │   ├── note
    │   │   ├── 2017-06-13-a-quick-note.md
    │   │   └── 2017-06-14-another-note.md
    │   └── post
    │       ├── 2015-07-23-lorem-ipsum.md
    │       └── 2016-02-14-hello-markdown.md
    ├── layouts
    │   └── partials
    │       └── foot_custom.html
    └── public
        └── ...
```

LICENSE.md and README.md are not required components of a theme, but you definitely should choose a license for your source code so that other people can properly use your code, and a README can be the brief documentation of your software.

The file archetypes/default.md defines the default template based on which users can create new posts. In this theme, default.md only provided empty YAML metadata:

```
---
---
```

The most important directories of a theme are layouts/ and static/. HTML templates are stored under layouts/, and assets are stored under static/.

To understand layouts/, you must know some basics about HTML (see Section B.1) because the templates under this directory are mostly HTML documents or fragments. There are many possible types of subdirectories under layouts/, but we are only going to introduce two here: _default/ and partials/.

- The _default/ directory is where you put the default templates for your web pages. In the XMin theme, we have three templates: single.html, list.html, and terms.html.

  - single.html is a template for rendering single pages. A single page basically corresponds to a Markdown document under content/, and it contains both the (YAML) metadata and content. Typically we want to render the page title, author, date, and the content. Below is the source code of XMin's single.html:

```
{{ partial "header.html" . }}
<div class="article-meta">
<h1><span class="title">{{ .Title }}</span></h1>
{{ with .Params.author }}
<h2 class="author">{{ . }}</h2>
{{ end }}
{{ if .Params.date }}
<h2 class="date">{{ .Date.Format "2006/01/02" }}</h2>
{{ end }}
</div>

<main>
{{ .Content }}
</main>

{{ partial "footer.html" . }}
```

You see a lot of pairs of double curly braces {{}}, and that is how you program the templates using Hugo's variables and functions.

The template starts with a partial template header.html, for which you will see the source code soon. For now, you can imagine it as all the HTML tags before the body of your page (e.g., <html><head>). Partial templates are mainly for reusing HTML code. For example, all HTML pages may share very similar <head></head> tags, and you can factor out the common parts into partial templates.

The metadata of a page is included in a <div> element with the class article-meta. We recommend that you assign classes to HTML elements when designing templates, so that it will be easier to apply CSS styles to these elements using class names. In a template, you have access to many variables provided by Hugo, e.g., the `.Title` variable stores the value of the page title, and we write the title in a `<span>` in a first-level header `<h1>`. Similarly, the author and date are written in `<h2>`, but only if they are provided in the YAML metadata. The syntax `{{ with FOO }}{{ . }}{{ end }}` is a shorthand of `{{if FOO }}{{ FOO }}{{ end }}`, i.e., it saves you the effort of typing the expression `FOO` twice by using `{{ . }}`. The method `.Format` can be applied to a date object, and in this theme, we format dates in the form `YYYY/mm/dd` (`2006/01/02` is the way to specify the format in Go).

Then we show the content of a page, which is stored in the variable `.Content`. The content is wrapped in a semantic HTML tag `<main>`.

The template is finished after we include another partial template `footer.html` (source code to be shown shortly).

To make it easier to understand how a template works, we show a minimal example post below:

```
---
title: Hello World
author: Frida Gomam
date: 2017-06-19
---

A single paragraph.
```

Using the template `single.html`, it will be converted to an HTML page with source code that looks more or less like this (with the header and footer omitted):

```
<div class="article-meta">
  <h1><span class="title">Hello World</span></h1>
  <h2 class="author">Frida Gomam</h2>
  <h2 class="date">2017/06/19</h2>
</div>
```

```
<main>
  <p>A single paragraph.</p>
</main>
```

For a full example of a single page, you may see `https://xmin.yihui.name/about/`.

– `list.html` is the template for rendering lists of pages, such as a list of blog posts, or a list of pages within a category or tag. Here is its source code:

```
{{ partial "header.html" . }}

{{if not .IsHome }}
<h1>{{ .Title }}</h1>
{{ end }}

{{ .Content }}

<ul>
  {{ range (where .Data.Pages "Section" "!=" "") }}
  <li>
    <span class="date">{{ .Date.Format "2006/01/02" }}</span>
    <a href="{{ .URL }}">{{ .Title }}</a>
  </li>
  {{ end }}
</ul>

{{ partial "footer.html" . }}
```

Again, it uses two partial templates `header.html` and `footer.html`. The expression `{{if not .IsHome }}` means, if this list is not the home page, show the page title. This is because I do not want to display the title on the homepage. It is just my personal preference. You can certainly display the title in `<h1>` on the home page if you want.

The `{{ .Content }}` shows the content of the list. Please note that

typically .Content is empty, which may be surprising. This is because a list page is not generated from a source Markdown file by
default. However, there is an exception. When you write a special
Markdown file _index.md under a directory corresponding to the
list name, the .Content of the list will be the content of this Markdown file. For example, you can define the content of your homepage
in content/_index.md, and the content of the post list page under
content/post/_index.md.

Next we generate the list using a loop (range) through all pages fil-
tered by the condition that the section of a page should not be empty.
"Section" in Hugo means the first-level subdirectory name under
content/. For example, the section of content/post/foo.md is post.
Therefore the filter means that we will list all pages under subdirectories of content/. This will exclude pages under the root content/
directory, such as content/about.md.

Please note that the variable .Data is dynamic, and its value changes
according to the specific list you want to generate. For example, the
list page https://xmin.yihui.name/post/ only contains pages under
content/post/, and https://xmin.yihui.name/note/ only contains
pages under content/note/. These list pages are automatically generated by Hugo, and you do not need to explicitly loop through the
sections post and note. That is, a single template list.html will generate multiple lists of pages according to the sections and taxonomy
terms (e.g., categories and tags) you have on your website.

The list items are represented by the HTML tags <li> in <ul>. Each
item consists of the date, link, and title of a page. You may see https:
//xmin.yihui.name/post/ for a full example of a list page.

- terms.html is the template for the home page of taxonomy terms. For
  example, you can use it to generate the full list of categories or tags.
  The source code is below:

```
{{ partial "header.html" . }}

<h1>{{ .Title }}</h1>
```

```
<ul class="terms">
  {{ range $key, $value := .Data.Terms }}
  <li>
    <a href='{{ (print "/" $.Data.Plural "/" $key) | relURL }}'>
      {{ $key }}
    </a>
    ({{ len $value }})
  </li>
  {{ end }}
</ul>


{{ partial "footer.html" . }}
```

Similar to list.html, it also uses a loop. The variable .Data.Terms
stores all terms under a taxonomy, e.g., all category names. You can
think of it as a named list in R (called a map in Go), with the names
being the terms and the values being lists of pages. The variable $key
denotes the term and $value denotes the list of pages associated with
this term. What we render in each <li> is a link to the term page as
well as the count of posts that used this term (len is a Go function
that returns the length of an object).

Hugo automatically renders all taxonomy pages, and the path names
are the plural forms of the taxonomies, e.g., https://xmin.yihui.
name/categories/ and https://xmin.yihui.name/tags/. That is the
meaning of .Data.Plural. The leading $ is required because we are
inside a loop, and need to access variables from the outside scope.
The link of the term is passed to the Hugo function relURL via a pipe
| to make it relative, which is good practice because relative links are
more portable (independent of the domain name).

- The partials/ directory is the place to put the HTML fragments to be
  reused by other templates via the partial function. We have four partial
  templates under this directory:

  – header.html main defines the <head> tag and the navigation menu
    in the <nav> tag.

```html
<!DOCTYPE html>
<html lang="{{ .Site.LanguageCode }}">
  <head>
    <meta charset="utf-8">
    <title>{{ .Title }} | {{ .Site.Title }}</title>
    <link href='{{ "/css/style.css" | relURL }}'
      rel="stylesheet" />
    <link href='{{ "/css/fonts.css" | relURL }}'
      rel="stylesheet" />
    {{ partial "head_custom.html" . }}
  </head>

  <body>
    <nav>
    <ul class="menu">
      {{ range .Site.Menus.main }}
      <li><a href="{{ .URL | relURL }}">{{ .Name }}</a></li>
      {{ end }}
    </ul>
    <hr/>
    </nav>
```

The `<head>` area should be easy to understand if you are familiar with HTML. Note that we also included a partial template `head_custom.html`, which is empty in this theme, but it will make it much easier for users to add customized code to `<head>` without rewriting the whole template. See Section 2.6 for more details.

The navigation menu is essentially a list, and each item of the list is read from the variable `.Site.Menus.main`. This means users can define the menu in `config.toml`, e.g.,

```toml
[[menu.main]]
    name = "Home"
    url = "/"
[[menu.main]]
```

```
    name = "About"
    url = "/about/"
```

It will generate a menu like this:

```html
<ul class="menu">
  <li><a href="/">Home</a></li>
  <li><a href="/about/">About</a></li>
</ul>
```

Hugo has a powerful menu system, and we only used the simplest type of menu in this theme. If you are interested in more features like nested menus, please see the full documentation at http://gohugo. io/extras/menus/.

– footer.html defines the footer area of a page and closes the HTML document:

```html
<footer>
{{ partial "foot_custom.html" . }}
{{ with .Site.Params.footer }}
<hr/>
{{ . | markdownify }}
{{ end }}
</footer>
</body>
</html>
```

The purpose of the partial template foot_custom.html is the same as head_custom.html; that is, to allow the user to add customized code to the <footer> without rewriting the whole template.

Lastly, we use the variable .Site.Params.footer to generate a page footer. Note we used the with function again. Recall that the syntax {{ with .Site.Params.footer }}{{ . }}{{ end }} is a short-hand for {{if .Site.Params.footer }}{{ .Site.Params.footer }}{{ end }}. This syntax saves you from typing the expression

.Site.Params.footer twice by using {{ . }} as a placeholder for the variable footer, which is defined as a site parameter in our config.toml file. The additional function markdownify can convert Markdown to HTML (i.e., {{ . | markdownify }}. Altogether, this sequence means we can define a footer option using Markdown under params in config.toml, e.g.,

```
[params]
    footer = "&copy; [Yihui Xie](https://yihui.name) 2017"
```

There is a special template 404.html, which Hugo uses to create the 404 page (when a page is not found, this page is displayed):

```
{{ partial "header.html" . }}

404 NOT FOUND

{{ partial "footer.html" . }}
```

With all templates above, we will be able to generate a website from Markdown source files. You are unlikely to be satisfied with the website, however, because the HTML elements are not styled at all, and the default appearance may not look appealing to most people. You may have noticed that in header.html, we have included two CSS files, /css/style.css and /css/fonts.css.

You can find many existing open-source CSS frameworks online that may be applied to a Hugo theme. For example, the most popular CSS framework may be Bootstrap: http://getbootstrap.com. When I was designing XMin, I wondered how far I could go without using any of these existing frameworks, because they are usually very big. For example, bootstrap.css has nearly 10000 lines of code when not minimized. It turned out that I was able to get a satisfactory appearance with about 50 lines of CSS, which I will explain in detail below:

- style.css defines all styles except the typefaces:

```
body {
    max-width: 800px;
    margin: auto;
    padding: 1em;
    line-height: 1.5em;
}
```

The maximum width of the page body is set to 800 pixels because an excessively wide page is difficult to read (800 is an arbitrary threshold that I picked). The body is centered using the CSS trick margin: auto, which means the top, right, bottom, and left margins are automatic. When a block element's left and right margins are auto, it will be centered.

```
/* header and footer areas */
.menu li { display: inline-block; }
.article-meta, .menu a {
    text-decoration: none;
    background: #eee;
    padding: 5px;
    border-radius: 5px;
}
.menu, .article-meta, footer { text-align: center; }
.title { font-size: 1.1em; }
footer a { text-decoration: none; }
hr {
    border-style: dashed;
    color: #ddd;
}
```

Remember that our menu element is a list <ul class="menu"> defined in header.html. I changed the default display style of <li> within the menu to inline-block, so that they will be laid out from left to right as inline elements, instead of being stacked vertically as a bullet list (the default behavior).

For links (<a>) in the menu and the metadata area of an article, the default text decoration (underlines) is removed, and a light background color is applied. The border radius is set to 5 pixels so that you can see a subtle round-corner rectangle behind each link.

The horizontal rule (<hr>) is set to a dashed light-gray line to make it less prominent on a page. These rules are used to separate the article body from the header and footer areas.

```css
/* code */
pre {
  border: 1px solid #ddd;
  box-shadow: 5px 5px 5px #eee;
  padding: 1em;
  overflow-x: auto;
}
code { background: #f9f9f9; }
pre code { background: none; }
```

For code blocks (<pre>), I apply light gray borders with drop-shadow effects. Every inline code element has a very light gray background. These decorations are merely out of my own peculiar interest and emphasis in code.

```css
/* misc elements */
img, iframe, video { max-width: 100%; }
main { hyphens: auto; }
blockquote {
  background: #f9f9f9;
  border-left: 5px solid #ccc;
  padding: 3px 1em 3px;
}

table {
  margin: auto;
  border-top: 1px solid #666;
  border-bottom: 1px solid #666;
}
table thead th { border-bottom: 1px solid #ddd; }
```

```
th, td { padding: 5px; }
tr:nth-child(even) { background: #eee }
```

Embedded elements like images and videos that exceed the page margin
are often ugly, so I restrict their maximum width to 100%. Hyphenation
is turned on for words in <main>. Blockquotes have a gray left sidebar and
a light gray background. Tables are centered by default, with only three
horizontal rules: the top and bottom borders of the table, and the bottom
border of the table head. Table rows are striped to make it easier to read
the table especially when the table is wide.

- fonts.css is a separate style sheet because it plays a critical role in the appearance of a website, and it is very likely that you will want to customize
  this file. In most cases, your readers will spend the most time on reading
  the text on your pages, so it is important to make the text comfortable to
  read. I'm not an expert in web design, and I just picked Palatino for the
  body and Lucida Console or Monaco (whichever is available in your system) for the code. It is common to use Google web fonts nowadays. You
  may try some web fonts and see if you like any of them.

```
body {
    font-family: "Palatino Linotype", "Book Antiqua", Palatino, serif;
}
code {
    font-family: "Lucida Console", Monaco, monospace;
    font-size: 85%;
}
```

The two CSS files are placed under the static/css/ directory of the theme.
In the HTML template header.html, the path /css/style.css refers to the
file static/css/style.css.

Lastly, this theme provided an example site under exampleSite/. The directory structure may be a little confusing because this is a theme instead of
a website. In practice, everything under exampleSite/ should be under the
root directory of a website, and the top-level hugo-xmin/ directory should be
under the themes/ directory of this website, i.e.,

```
├── config.toml
├── content/
├── ...
├── themes/
│    └── hugo-xmin/
│
└── ...
```

The example site provides a sample `config.toml`, a home page `_index.md`, an about page `about.md`, two posts under `note/` and two under `post/`. It also overrides the `foot_custom.html` in the theme.

2.5.2   Implementing more features

The XMin is actually a highly functional theme, but we understand that it may be too minimal for you. There are a few commonly used features (intentionally) missing in this theme, and we will teach you how to add them by yourself if desired. All these features and the source code can be applied to other themes, too.

- **Enable Google Analytics.** Hugo has provided a built-in partial template. For XMin, you can add

```
{{ template "_internal/google_analytics.html" . }}
```

  to `layouts/partials/foot_custom.html` under the root directory of your website (instead of `themes/hugo-xmin/`), and configure `googleAnalytics` in the `config.toml`. See `https://github.com/yihui/hugo-xmin/pull/3` for details, and the HTML source of this page for the JavaScript rendered from the template: `https://deploy-preview-3--hugo-xmin.netlify.com`.

- **Enable Disqus comments.** Similar to Google Analytics, you can add the built-in template

```
{{ template "_internal/disqus.html" . }}
```

to `foot_custom.html`, and configure the Disqus shortname in `config.toml`. See `https://github.com/yihui/hugo-xmin/pull/4` for details, and a preview at `https://deploy-preview-4--hugo-xmin.netlify.com`.

- **Enable syntax highlighting via highlight.js.** Add this to `head_custom.html`

```
<link href="//YOUR-CDN-LINK/styles/github.min.css" rel="stylesheet">
```

and this to `foot_custom.html`:

```
<script src="//YOUR-CDN-LINK/highlight.min.js"></script>
<script src="//YOUR-CDN-LINK/languages/r.min.js"></script>

<script>
hljs.configure({languages: []});
hljs.initHighlightingOnLoad();
</script>
```

Remember to replace `YOUR-CDN-LINK` with the link to your preferred CDN host of highlight.js, e.g., `cdn.bootcss.com/highlight.js/9.12.0`. For more information about highlight.js, please see its homepage: `https://highlightjs.org`. If you need to use other CDN hosts, cdnjs.com is a good choice: `https://cdnjs.com/libraries/highlight.js` You can also see which languages and CSS themes are supported there.

You may see `https://github.com/yihui/hugo-xmin/pull/5` for an actual implementation, and a sample page with syntax highlighting at `https://deploy-preview-5--hugo-xmin.netlify.com/post/2016/02/14/a-plain-markdown-post/`.

- **Support math expressions through MathJax.** Add the code below to `foot_custom.html`.

```
<script src="//yihui.name/js/math-code.js"></script>
<script async
src="//cdn.bootcss.com/mathjax/2.7.1/MathJax.js?config=TeX-MML-AM_CHTML">
</script>
```

This requires substantial knowledge of JavaScript and familiarity with MathJax to fully understand the code above, and we will leave the explanation of the code to Section B.3.

Note that bootcss.com is only one possible CDN host of MathJax, and you are free to use other hosts.

- **Show the table of contents (TOC).** To show a TOC for R Markdown posts, you only need to add the output format `blogdown::html_page` with the option `toc: true` to YAML:

```
output:
  blogdown::html_page:
    toc: true
```

For plain Markdown posts, you have to modify the template `single.html`. The TOC of a post is stored in the Hugo template variable `.TableOfContents`. You may want an option to control whether to show the TOC, e.g., you may add an option `toc: true` to the YAML metadata of a Markdown post to show the TOC. The code below can be added before the content of a post in `single.html`:

```
{{ if .Params.toc }}
{{ .TableOfContents }}
{{ end }}
```

See `https://github.com/yihui/hugo-xmin/pull/7` for an implementation with examples.

- **Display categories and tags in a post if provided in its YAML.** Add the code below where you want to place the categories and tags in `single.html`, e.g., in `<div class="article-meta"></div>`.

```
<p class="terms">
  {{ range $i := (slice "categories" "tags") }}
  {{ with ($.Param $i) }}
  {{ $i | title }}:
  {{ range $k := . }}
```

```
<a href='{{ relURL (print "/" $i "/" $k | urlize) }}'>{{$k}}</a>
{{ end }}
{{ end }}
{{ end }}
</p>
```

Basically the code loops through the YAML metadata fields `categories` and `tags`, and for each field, its value is obtained from `.Param`, then we use an inside loop to write out the terms with links of the form `<a href="/tags/foo/">foo</a>`.

You may see `https://github.com/yihui/hugo-xmin/pull/2` for the complete implementation and a preview at `https://deploy-preview-2--hugo-xmin.netlify.com/post/2016/02/14/a-plain-markdown-post/`.

- **Add pagination.** When you have a large number of posts on a website, you may not want to display the full list on a single page, but show N posts (e.g., N = 10) per page. It is easy to add pagination to a website using Hugo's built-in functions and templates. Instead of looping through all posts in a list template (e.g., `range .Data.Pages`), you paginate the full list of posts using the function `.Paginate` (e.g., `range (.Paginate .Data.Pages)`). Below is a template fragment that you may insert to your template file `list.html`:

```
<ul>
  {{ $paginator := .Paginate .Data.Pages }}
  {{ range $paginator.Pages }}
  <li>
    <span class="date">{{ .Date.Format "2006/01/02" }}</span>
    <a href="{{ .URL }}">{{ .Title }}</a>
  </li>
  {{ end }}
</ul>
{{ template "_internal/pagination.html" . }}
```

See `https://github.com/yihui/hugo-xmin/pull/16` for a full implementation.

- **Add a GitHub Edit button or link to a page.** If none of the above features look exciting to you (which would not surprise me), this little feature is really a great example of showing you the power of plain-text files and static websites, when combined with GitHub (or other services that support the online editing of plain-text files). I believe it would be difficult, if not impossible, to implement this feature in dynamic website frameworks like WordPress.

Basically, when you browse any text files in a repository on GitHub, you can edit them right on the page by hitting the Edit button (see Figure 2.3 for an example) if you have a GitHub account. If you have write access to the repository, you can commit the changes directly online, otherwise GitHub will fork the repository for you automatically, so that you can edit the file in your own repository, and GitHub will guide you to create a pull request to the original repository. When the original owner sees the pull request, he/she can see the changes you made and decide whether to accept them or not or ask you to make further changes. Although the terminology "pull request" is highly confusing to beginners,[18] it is probably the single greatest feature invented by GitHub, because it makes it so much easier for people to make contributions.

What is really handy is that all you need is a URL of a fixed form to edit a file on GitHub: `https://github.com/USER/REPO/edit/BRANCH/PATH/TO/FILE`. For example, `https://github.com/rbind/yihui/edit/master/content/knitr/faq.md`, where `USER` is `rbind`, `REPO` is `yihui`, `BRANCH` is `master`, and the file path is `content/knitr/faq.md`.

The key to implementing this feature is the variable `.File.Path`, which gives us the source file path of a page under `content/`, e.g., `post/foo.md`. If your website only uses plain Markdown files, the implementation will be very simple. I omitted the full GitHub URL in ... below, of which an example could be `https://github.com/rbind/yihui/edit/master/content/`.

```
{{ with .File.Path }}
<a href="https://github.com/.../{{ . }}">Edit this page</a>
{{ end }}
```

However, the case is a little more complicated for **blogdown** users, when

---

[18] In my opinion, it really should be called "merge request" instead.

R Markdown posts are involved. You cannot just use .File.Path because it actually points to the .html output file from an .Rmd file, whereas the .Rmd file is the actual source file. The Edit button or link should not point to the .html file. Below is the complete implementation that you may add to a template file depending on where you want to show the Edit link (e.g., footer.html):

```
{{ if .File.Path }}

{{ $Rmd := (print .File.BaseFileName ".Rmd")) }}

{{ if (where (readDir (print "content/" .File.Dir)) "Name" $Rmd) }}
  {{ $.Scratch.Set "FilePath" (print .File.Dir $Rmd) }}
{{ else }}
  {{ $.Scratch.Set "FilePath" .File.Path }}
{{ end }}

{{ with .Site.Params.GithubEdit}}
<a href='{{ . }}{{ $.Scratch.Get "FilePath" }}'>Edit this page</a>
{{ end }}

{{ end }}
```

The basic logic is that for a file, if the same filename with the extension .Rmd exists, we will point the Edit link to the Rmd file. First, we define a variable $Rmd to be the filename with the .Rmd extension. Then we check if it exists. Unfortunately, there is no function in Hugo like file.exists() in R, so we have to use a hack: list all files under the directory and see if the Rmd file is in the list. $.Scratch[19] is the way to dynamically store and obtain variables in Hugo templates. Most variables in Hugo are read-only, and you have to use $.Scratch when you want to modify a variable. We set a variable FilePath in $.Scratch, whose value is the full path to the Rmd file when the Rmd file exists, and the path to the Markdown source file otherwise. Finally, we concatenate a custom option GithubEdit in config.toml with the file path to complete the Edit link <a>. Here is an example of the option in config.toml:

---

[19]http://gohugo.io/extras/scratch/

```
[params]
  GithubEdit = "https://github.com/rbind/yihui/edit/master/content/"
```

Please note that if you use Hugo on Windows to build and deploy your site, you may have to change the file path separators from backslashes to forward slashes, e.g., you may need `{{ $.Scratch.Set "FilePath" (replace ($.Scratch.Get "FilePath") "\\" "/") }}` in the template. To avoid this complication, we do not recommend that you deploy your site through Windows (see Chapter 3 for deployment methods).

You may see `https://github.com/yihui/hugo-xmin/pull/6` for an actual implementation with R Markdown examples, and see the footer of this page for the Edit link: `https://deploy-preview-6--hugo-xmin.netlify.com`. You can actually see a link in the footer of every page, except the lists of pages (because they do not have source files).

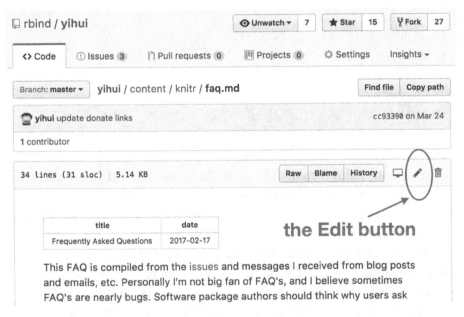

FIGURE 2.3: Edit a text file online on GitHub.

After you digest the XMin theme and the implementations of additional features, it should be much easier to understand other people's templates. There are a large number of Hugo themes but the primary differences among them are often in styles. The basic components of templates are often similar.

2.6 Custom layouts

It is very likely that you want to customize a theme unless you designed it. The most straightforward way is to simply make changes directly in the theme,[20] but the problem is that a Hugo theme may be constantly updated by its original author for improvements or bug fixes. Similar to the "you break it, you buy it" policy (the Pottery Barn rule[21]), once you touch someone else's source code, you will be responsible for its future maintenance, and the original author should not be responsible for the changes you made on your side. That means it may not be easy to pull future updates of this theme to your website (you have to carefully read the changes and make sure they do not conflict with your changes), but if you are completely satisfied with the current state of the theme and do not want future updates, it is fine to modify the theme files directly.

A theme author who is aware of the fact that users may customize her theme will typically provide two ways: one is to provide options in `config.toml`, so that you can change these options without touching the template files; the other is to leave a few lightweight template files under `layouts/` in the theme, so that you can override them without touching the core template files. Take the XMin theme for example:

I have two empty HTML files `head_custom.html` and `foot_custom.html` under `layouts/partials/` in the theme. The former will be added inside `<head></head>` of a page, e.g., you can load JavaScript libraries or include CSS style sheets via `<link>`. The latter will be added before the footer of a page, e.g., you may load additional JavaScript libraries or embed Disqus comments there.

---

[20] If you are new to web development, be careful changing content within the theme. Minor changes like colors and font sizes can be found within the CSS files of the theme and can be altered simply with minimal risk of breaking the theme's functionality.

[21] https://en.wikipedia.org/wiki/Pottery_Barn_rule

The way that you customize these two files is not to edit them directly in
the theme folder, but to create a directory `layouts/partials/` under the root
directory of your website, e.g., your directory structure may look like this:

```
your-website/
├── config.toml
├── ...
├── themes/
│   └── hugo-xmin/
│       ├── ...
│       └── layouts/
│           ├── ...
│           └── partials
│               ├── foot_custom.html
│               ├── footer.html
│               ├── head_custom.html
│               └── header.html
└── layouts
    └── partials
        ├── foot_custom.html
        └── head_custom.html
```

All files under `layouts/` under the root directory will override files with
the same relative paths under `themes/hugo-xmin/layouts/`, e.g., the
file `layouts/partials/foot_custom.html`, when provided, will override
`themes/hugo-xmin/layouts/partials/foot_custom.html`.   That   means
you only need to create and maintain at most two files under `layouts/`
instead of maintaining all files under `themes/`. Note that this overriding
mechanism applies to all files under `layouts/`, and is not limited to the
`partials/` directory. It also applies to any Hugo theme that you actually use
for your website, and is not limited to `hugo-xmin`.

2.7 Static files

All files under the `static/` directory are copied to `public/` when Hugo renders a website. This directory is often used to store static web assets like images, CSS, and JavaScript files. For example, an image `static/foo/bar.png` can be embedded in your post using the Markdown syntax `![](/foo/bar.png)`.[22]

Usually a theme has a `static/` folder, and you can partially override its files using the same mechanism as overriding `layouts/` files, i.e., `static/file` will override `themes/theme-name/static/file`. In the XMin theme, I have two CSS files `style.css` and `fonts.css`. The former is the main style sheet, and the latter is a quite small file to define typefaces only. You may want to define your own typefaces, and you can only provide a `static/css/fonts.css` to override the one in the theme, e.g.,

```
body {
  font-family: "Comic Sans MS", cursive, sans-serif;
}
code {
  font-family: "Courier New", Courier, monospace;
}
```

To R Markdown users, another important application of the `static/` directory is to build Rmd documents with custom output formats, i.e., Rmd documents not using the `blogdown::html_page()` format (see Section 1.5). For example, you can generate a PDF or presentations from Rmd documents under this directory, so that Hugo will not post-process them but simply copies them to `public/` for publishing. To build these Rmd files, you must provide a custom build script `R/build.R` (see Section D.9). You can write a single line of code in this script:

---

[22]The link of the image depends on your `baseurl` setting in `config.toml`. If it does not contain a subpath, `/foo/bar.png` will be the link of the image, otherwise you may have to adjust it, e.g., for `baseurl = "http://example.com/subpath/"`, the link to the image should be `/subpath/foo/bar.png`.

```
blogdown::build_dir("static")
```

The function `build_dir()` finds all Rmd files under a directory, and calls `rmarkdown::render()` to build them to the output formats specified in the YAML metadata of the Rmd files. If your Rmd files should not be rendered by a simple `rmarkdown::render()` call, you are free to provide your own code to render them in `R/build.R`. There is a built-in caching mechanism in the function `build_dir()`: an Rmd file will not be compiled if it is older than its output file(s). If you do not want this behavior, you can force all Rmd files to be recompiled every time: `build_dir(force = TRUE)`.

I have provided a minimal example in the GitHub repository yihui/blogdown-static,[23] where you can find two Rmd examples under the `static/` directory. One is an HTML5 presentation based on the **xaringan** package, and the other is a PDF document based on **bookdown**.

You need to be cautious about arbitrary files under `static/`, due to Hugo's overriding mechanism. That is, everything under `static/` will be copied to `public/`. You need to make sure that the files you render under `static/` will not conflict with those files automatically generated by Hugo from `content/`. For example, if you have a source file `content/about.md` and an Rmd file `static/about/index.Rmd` at the same time, the HTML output from the latter will overwrite the former (both Hugo and you will generate an output file with the same name `public/about/index.html`).

---

[23] https://github.com/yihui/blogdown-static

3

Deployment

Since the website is basically a folder containing static files, it is much easier to deploy than websites that require dynamic server-side languages such as PHP or databases. All you need is to upload the files to a server, and usually your website will be up and running shortly. The key question is which web server you want to use. If you do not have your own server, you may try the ones listed in this chapter. Most of them are free (except Amazon S3), or at least provide free plans. Disclaimer: the authors of this book are not affiliated with any of these services or companies, and there is no guarantee that these services will be provided forever.[1]

Considering the cost and friendliness to beginners, we currently recommend Netlify (`https://www.netlify.com`). It provides a free plan that actually has quite a lot of useful features. If you have no experience in publishing websites before, just log in using your GitHub account or other accounts, drag the `public/` folder built by **blogdown** for your website to the Netlify page, and your website will be online in a few seconds with a random subdomain name of the form `random-word-12345.netlify.com` provided by Netlify (you can customize the name). You can easily automate this process (see Section 3.1 for more details). You do not need to wrestle with `ssh` or `rsync -zrvce` anymore, if you know what these commands mean.

The second easiest solution may be Updog (`https://updog.co`), which features Dropbox integration. Publishing a website can be as easy as copying the files under the `public/` folder of your **blogdown** website to a Dropbox folder. The free plan of Updog only provides limited features, and its paid plan will give you access to much richer features.

If you do not mind using command-line tools or are familiar with GIT/GitHub, you can certainly consider services like GitHub Pages, Travis CI, or Amazon S3 to build or host your websites. No matter which service

---

[1]You can easily find other similar services if you use your search engine.

you use, please keep in mind that none of them can really lock you in and you are always free to change the service. As we have mentioned before, one great advantage of **blogdown** is that your website will be a folder of static files that you can move to any web server.

3.1 Netlify

As we just mentioned, Netlify allows you to quickly publish a website by uploading the `public/` folder through its web interface, and you will be assigned a random subdomain `*.netlify.com`.[2] This approach is good for those websites that are not updated frequently (or at all). However, it is unlikely that you will not need to update your website, so we introduce a better approach in this section,[3] which will take you a few more minutes to complete the configurations. Once it is properly configured, all you need to do in the future is to update the source repository, and Netlify will call Hugo to render your website automatically.

Basically, you have to host all source files of your website in a GIT repository.[4] You do not need to put the `public/` directory under version control[5] because it will be automatically generated. Currently, Netlify supports GIT repositories hosted on GitHub, GitLab, and BitBucket. With any of these accounts, you can log into Netlify from its homepage and follow the guide to create a new site from your GIT repository.

Netlify supports several static website generators, including Jekyll and Hugo. For a new site, you have to specify a command to build your website, as well as the path of the publish directory. Netlify also supports multiple ver-

---

[2] You don't have to keep the `*.netlify.com` domain. See Appendix C for more information.

[3] Please bear in mind that the purpose of this section is to outline the basic steps of publishing a website with Netlify, and the technical details may change from time to time, so the official Netlify documentation should be the most reliable source if you have any questions or anything we introduced here does not work.

[4] If the contents of your `blogdown` site are not in the root directory of your GIT repository, Netlify will not build.

[5] You can add `public` to `.gitignore` to ignore it in GIT.

sions of Hugo, so the build command can be the default hugo. The default version is 0.17, which is too old, and we recommend that you use at least version 0.20. To specify a Hugo version greater or equal to 0.20, you need to create an environment variable HUGO_VERSION on Netlify. See the Netlify documentation[6] for more information. The publish directory should be public unless you have changed it in your config.toml. Figure 3.1 shows the settings of the website https://t.yihui.name. You do not have to follow the exact settings for your own website; in particular, you may need to change the value of the environment variable HUGO_VERSION to a recent version of Hugo.[7]

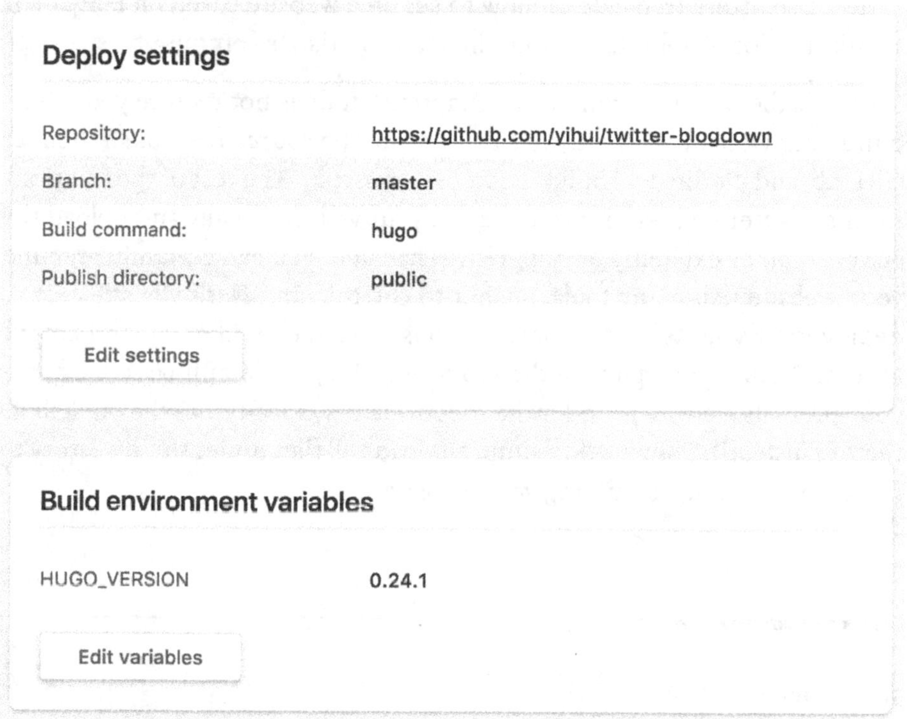

Deploy settings

Repository:	https://github.com/yihui/twitter-blogdown
Branch:	master
Build command:	hugo
Publish directory:	public

Edit settings

Build environment variables

HUGO_VERSION	0.24.1

Edit variables

FIGURE 3.1: Example settings of a website deployed on Netlify.

It may take a minute or two to deploy your website on Netlify for the first time, but it can be much faster later (a few seconds) when you update your website source, because Netlify deploys incremental changes in the public/ directory, i.e., only the newer files compared to the last time are deployed.

---

[6]https://www.netlify.com/docs/continuous-deployment/
[7]By the time when this book is published, the version 0.24.1 may be too old.

After your GIT repository is connected with Netlify, the last issue you may want to solve is the domain name, unless you are satisfied with the free Netlify subdomain. If you want to use a different domain, you need to configure some DNS records of the domain to point it to the Netlify server. See Appendix C for some background knowledge on domain names.

If you are not familiar with domain names or do not want to learn more about them, something to consider is a free subdomain `*.rbind.io` offered by RStudio, Inc. Please visit the Rbind support website `https://support.rbind.io` to learn how to apply for a subdomain. In fact, the Rbind organization also offers free help on how to set up a website based on **blogdown**, thanks to a lot of volunteers from the R and statistics community.

Netlify is the only solution in this chapter that does not require you to prebuild your website. You only need to update the source files, push them to GitHub, and Netlify will build the website for you.[8] The rest of the solutions in this chapter will require you to build your website locally and upload the `public/` folder explicitly or implicitly. That said, you can certainly prebuild your website using any tools, push it to GitHub, and Netlify is still able to deploy it for you. What you need to do is leave the build command empty, and tell Netlify your publish directory (e.g., Hugo's default `public/`, but if your prebuilt website is under the root directory, specify . as the publish directory instead). Then Netlify simply uploads all files under this directory to its servers without rebuilding your website.

3.2   Updog

Updog (`https://updog.co`) provides a simple service: it turns a specified Dropbox (or Google Drive) folder into a website. The idea is that you grant Updog the permission to read the folder, and it will act as a middleman to serve your files under this folder to your visitors. This folder has to be accessed via a domain name, and Updog offers a free subdomain `*.updog.co`. For example, if you have assigned the domain `example.updog.co` to your Dropbox folder, and a visitor wants to see the

---

[8]This is called "continuous deployment."

page `https://example.updog.co/foo/index.html`, Updog will read the file `foo/index.html` in your Dropbox folder and display it to the visitor.

At the moment, Updog's free plan only allows one website per account and will insert a footer "Hosted on Updog" on your web pages. You may not like these limitations. The major benefit of using Updog is that publishing a website becomes implicit since Dropbox will continuously sync files. All you need to do is to make sure your website is generated to the correct Dropbox folder. This can be easily achieved by setting the option `publishDir` in `config.toml`. For example, suppose the folder that you assign to Updog is `~/Dropbox/Apps/updog/my-website/`, and your source folder is at `~/Dropbox/Apps/updog/my-source/`, then you can set `publishDir: "../my-website"` in `~/Dropbox/Apps/updog/my-source/config.toml`.

You can also use your custom domain name if you do not want the default Updog subdomain, and you only need to point the CNAME record of your domain name to the Updog subdomain.[9]

3.3 GitHub Pages

GitHub Pages (`https://pages.github.com`) is a very popular way to host static websites (especially those built with Jekyll), but its advantages are not obvious or appealing compared to Netlify. We recommend that you consider Netlify + Hugo due to these reasons:

- Currently, GitHub Pages does not support HTTPS for custom domain names. HTTPS only works for `*.github.io` subdomains. This limitation does not exist on Netlify. You may read the article "Why HTTPS for Everything?"[10] to understand why it is important, and you are encouraged to turn on HTTPS for your website whenever you can.

- Redirecting URLs is awkward with GitHub Pages but much more straightforward with Netlify.[11] This is important especially when you have an old

---

[9]See Appendix C for more information.

[10]`https://https.cio.gov/everything/`

[11]GitHub Pages uses a Jekyll plugin to write an HTTP-REFRESH meta tag to redirect pages,

website that you want to migrate to Hugo; some links may be broken, in which case you can easily redirect them with Netlify.

- One of the best features of Netlify that is not available with GitHub Pages is that Netlify can generate a unique website for preview when a GitHub pull request is submitted to your GitHub repository. This is extremely useful when someone else (or even yourself) proposes changes to your website, since you have a chance to see what the website would look like before you merge the pull request.

Basically, Netlify can do everything that GitHub Pages can, but there is still one little missing feature, which is closely tied to GitHub itself, which is GitHub Project Pages.[12] This feature allows you to have project websites in separate repositories, e.g., you may have two independent websites `https://username.github.io/proj-a/` and `https://username.github.io/proj-b/`, corresponding to GitHub repositories `username/proj-a` and `username/proj-b`, respectively. However, since you can connect any GitHub repositories with Netlify, and each repository can be associated with a domain or subdomain name, you may replace GitHub Project Pages with different subdomains like `proj-a.netlify.com` and `proj-b.netlify.com`. The actual limitation is that you cannot use subpaths in the URL but you can use any (sub)domain names.

Although GitHub does not officially support Hugo (only Jekyll is supported), you can actually publish any static HTML files on GitHub Pages, even if they are not built with Jekyll. The first requirement for using GitHub Pages is that you have to create a GitHub repository named `username.github.io` under your account (replace `username` with your actual GitHub username), and what's left is to push your website files to this repository. The comprehensive documentation of GitHub Pages is at `https://pages.github.com`, and please ignore anything related to Jekyll there unless you actually use Jekyll instead of Hugo. To make sure GitHub does not rebuild your website using Jekyll and just publish whatever files you push to the repository, you need to create a (hidden) file named `.nojekyll` in the repository.[13] GitHub offers

---

and Netlify can do pattern-based 301 or 302 redirects, which can notify search engines that certain pages have been moved (permanently or temporarily).

[12]`https://help.github.com/articles/user-organization-and-project-pages/`

[13]You may use the R function `file.create('.nojekyll')` to create this file if you do not know how to do this.

a free subdomain `username.github.io`, and you can use your own domain name by configuring its A or CNAME records to point it to GitHub Pages (consult the GitHub Pages documentation for instructions).

Your `public/` directory should be the GIT repository. You have two possible choices for setting up this repository locally. The first choice is to follow the default structure of a Hugo website like the diagram below, and initialize the GIT repository under the `public/` directory:

```
source/
|
├── config.toml
├── content/
├── themes/
├── ...
└── public/
    |
    ├── .git/
    ├── .nojekyll
    ├── index.html
    ├── about/
    └── ...
```

If you know how to use the command line, change the working directory to `public/`, and initialize the GIT repository there:

```
cd public
git init
git remote add origin https://github.com/username/username.github.io
```

The other choice is to clone the GitHub repository you created to the same directory as your website source:

```
git clone https://github.com/username/username.github.io
```

And the structure looks like this:

```
source/
|
├── config.toml
├── content/
├── themes/
└── ...


username.github.io/
|
├── .git/
├── .nojekyll
├── index.html
├── about/
└── ...
```

The source directory and the `username.github.io` directory are under the
same parent directory. In this case, you need to set the option `publishDir:`
`"../username.github.io"` in source/config.toml.

3.4   Travis + GitHub

If you decide not to follow our recommendation to use Netlify to deploy your
website, we should warn you that the approach in this section will require
substantial knowledge about GIT, GitHub, Travis CI (`https://travis-ci.`
`org`), and the Linux command line, which we will leave for you to learn on
your own. The major advantage of publishing via Travis CI is that you can
compile all your Rmd posts on Travis CI (on the cloud) instead of your local
computer.

In case you are not familiar with Travis, it is a service to continuously check
your software in a virtual machine whenever you push changes to GitHub.
It is primarily for testing software, but since you can run a lot of commands
in its virtual machine, you can certainly use the virtual machine to do other

things, e.g., install R and the **blogdown** package to build websites. Before I show you how, I'd like to mention two issues that you should be aware of:

- Personally, I prefer taking a look at the output in GIT to see the changes when I have any output that is dynamically computed from R, so that I know for sure what I'm going to publish exactly. With Travis, it is somewhat unpredictable because it is fully automatic and you do not have a chance to see the new content or results to be published. There are many factors that could affect building the site: the R version, availability of certain R packages, system dependencies, and network connection, etc.

- The time required to compile all Rmd files may be very long and cause timeouts on Travis, depending on how time-consuming your R code is. There is a caching mechanism in **blogdown** to speed up the building of your site (see Section D.9), and if you use Travis to build your website, you will not benefit from this caching mechanism unless you take advantage of Travis's caching. You have to cache the directories content/, static/, and blogdown/, but Travis's cache is a little fragile in my experience. Sometimes the cache may be purged for unknown reasons. What is more, you cannot directly cache content/ and static/, because Travis clones your repository before restoring the cache, which means old files from the cached content/ and static/ may overwrite new files you pushed to GitHub.

The second problem can be solved, but I do not want to explain how in this book since the solution is too involved. If you really want to use Travis to build your website and run into this problem, you may file an issue to the GitHub repository https://github.com/yihui/travis-blogdown. In fact, this repository is a minimal example I created to show how to build a website on Travis and publish to GitHub Pages.

The Travis documentation shows how to deploy a site to GitHub Pages: https://docs.travis-ci.com/user/deployment/pages/, but does not show how to build a site. Here is the Travis configuration file, .travis.yml, for the travis-blogdown repository:

```
language: r
dist: trusty
sudo: false
```

```
branches:
  only:
    - master

cache:
  packages: yes
  directories:
    - $HOME/bin

before_script:
  - "Rscript -e 'blogdown::install_hugo()'"

script:
  - "Rscript -e 'blogdown::build_site()'"

deploy:
  provider: pages
  skip_cleanup: true
  github_token: $GITHUB_TOKEN
  on:
    branch: master
  local_dir: public
  fqdn: travis-blogdown.yihui.name
```

The key is that we install Hugo via `blogdown::install_hugo()` and build the site via `blogdown::build_site()`. To trick Travis into building this repository like an R package, you must have a DESCRIPTION file in the repository, otherwise your website will not be built.

```
Package: placeholder
Type: Website
Title: Does not matter.
Version: 0.0.1
Imports: blogdown
Remotes: rstudio/blogdown
```

There are a few more things to explain and emphasize in .travis.yml:

- The branches option specifies that only changes in the master branch will trigger building on Travis.

- The cache option specifies all R packages to be cached, so the next time it will be faster to build the site (R packages do not need to be reinstalled from source). The bin/ directory in the home directory is also cached because Hugo is installed there, and the next time Hugo does not need to be reinstalled.

- For the deploy option, there is an environment variable named GITHUB_TOKEN, and I have specified its value to be a GitHub personal access token via the Travis settings of this repository, so that Travis will be able to write to my repository after the website is built. The option on specifies that the deployment will only occur when the master branch is built. The local_dir option is the publish directory, which should default to public in Hugo. By default, the website is pushed to the gh-pages branch of this repository. The fqdn option specifies the custom domain of the website. I have set a CNAME record (see Appendix C) to point travis-blogdown.yihui.name to yihui.github.io, so that GitHub is able to serve this website through this domain (in fact, Travis will write a CNAME file containing the domain to the gh-pages branch).

If you use the username.github.io repository on GitHub, the website must be pushed to its master branch instead of gh-pages (this is the only exception). I recommend that you separate the source repository and the output repository. For example, you may have a website-source repository with the same settings as the above .travis.yml except for two new options under deploy:

```
deploy:
  ...
  repo: username/username.github.io
  target_branch: master
```

This means the website will be pushed to the master branch of the repository username/username.github.io (remember to replace username with your actual username).

You can also deploy your website to Amazon S3, and the setup on the R side is very similar to what we have introduced for GitHub Pages. The only difference is in the last step, where you change the target from GitHub Pages to Amazon S3. For more information, please see the documentation on Travis: `https://docs.travis-ci.com/user/deployment/s3/`.

3.5   GitLab Pages

GitLab (`http://gitlab.com`) is a very popular way to host the source code of your project. GitLab has a built in Continuous Integration & Deployment (CI/CD) service[14] that can be used to host static websites, named GitLab Pages[15]. The major advantage of using GitLab Pages is that you will be able to compile all your Rmd posts through its CI/CD service instead of your local computer and any generated content, such as HTML files, will be automatically copied to the web server. Please note that this approach has similar issues as the Travis + GitHub approach in Section 3.4.

GitLab's CI/CD service uses the instructions stored in the YAML file `.gitlab-ci.yml` in the repository. Here is a sample configuration file `.gitlab-ci.yml` from the example repository `https://gitlab.com/rgaiacs/blogdown-gitlab`:

```
image: debian:buster-slim

before_script:
  - apt-get update && apt-get -y install pandoc r-base
  - R -e "install.packages('blogdown',repos='http://cran.rstudio.com')"
  - R -e "blogdown::install_hugo()"

pages:
  script:
    - R -e "blogdown::build_site()"
```

----

[14]`https://about.gitlab.com/features/gitlab-ci-cd/`
[15]`https://about.gitlab.com/features/pages/`

```
artifacts:
  paths:
    - public
only:
  - master
```

The `image` option specifies what Docker[16] image will be use as a start point. We are using a Debian image but any image from Docker Hub[17] can be used. Other settings and options are similar to `.travis.yml` in Section 3.4. The above example generates the website at `https://rgaiacs.gitlab.io/blogdown-gitlab`.

---

[16]`https://www.docker.com`
[17]`https://hub.docker.com/`

4

Migration

Usually, it is easier to start a new website than migrating an old one to a new framework, but you may have to do it anyway because of the useful content on the old website that should not simply be discarded. A lazy solution is to leave the old website as is, start a new website with a new domain, and provide a link to the old website. This may be a hassle to your readers, and they may not be able to easily discover the gems that you created on your old website, so I recommend that you migrate your old posts and pages to the new website if possible.

This process may be easy or hard, depending on how complicated your old website is. The bad news is that there is unlikely to be a universal or magical solution, but I have provided some helper functions in **blogdown** as well as a Shiny application to assist you, which may make it a little easier for you to migrate from Jekyll and WordPress sites.

To give you an idea about the possible amount of work required, I can tell you that it took me a whole week (from the morning to midnight every day) to migrate several of my personal Jekyll-based websites to Hugo and **blogdown**. The complication in my case was not only Jekyll, but also the fact that I built several separate Jekyll websites (because I did not have a choice in Jekyll) and I wanted to unite them in the same repository. Now my two blogs (Chinese and English), the **knitr** (Xie, 2017c) package documentation, and the **animation** package (Xie, 2017a) documentation are maintained in the same repository: `https://github.com/rbind/yihui`. I have about 1000 pages on this website, most of which are blog posts. It used to take me more than 30 seconds to preview my blog in Jekyll, and now it takes less than 2 seconds to build the site in Hugo.

Another complicated example is the website of Rob J Hyndman (`https://robjhyndman.com`). He started his website in 1993 (12 years before me), and had accumulated a lot of content over the years. You can read the post `https://support.rbind.io/2017/05/15/converting-robjhyndman-to-blogdown/` for the stories about how he migrated his WordPress website to **blogdown**. The key is that you probably need a long international flight when you want to migrate a complicated website.

A simpler example is the Simply Statistics blog (`https://simplystatistics.org`). Originally it was built on Jekyll[1] and the source was hosted in the GitHub repository `https://github.com/simplystats/simplystats.github.io`. I volunteered to help them move to **blogdown**, and it took me about four hours. My time was mostly spent on cleaning up the YAML metadata of posts and tweaking the Hugo theme. They had about 1000 posts, which sounds like a lot, but the number does not really matter, because I wrote an R script to process all posts automatically. The new repository is at `https://github.com/rbind/simplystats`.

If you do not really have too many pages (e.g., under 20), I recommend that you cut and paste them to Markdown files, because it may actually take longer to write a script to process these pages.

It is likely that some links will be broken after the migration because Hugo renders different links for your pages and posts. In that case, you may either fix the permanent links (e.g., by tweaking the slug of a post), or use 301 redirects (e.g., on Netlify).

4.1   From Jekyll

When converting a Jekyll website to Hugo, the most challenging part is the theme. If you want to keep exactly the same theme, you will have to rewrite your Jekyll templates using Hugo's syntax (see Section 2.5). However, if you can find an existing theme in Hugo (`https://themes.gohugo.io`), things will be much easier, and you only need to move the content of your website to Hugo, which is relatively easy. Basically, you copy the Markdown pages and posts to the `content/` directory in Hugo and tweak these text files.

---

[1]It was migrated from WordPress a few years ago. The WordPress site was actually migrated from an earlier Tumblr blog.

Usually, posts in Jekyll are under the _posts/ directory, and you can move them to content/post/ (you are free to use other directories). Then you need to define a custom rule for permanent URLs in config.toml like (see Section 2.2.2):

```
[permalinks]
    post = "/:year/:month/:day/:slug/"
```

This depends on the format of the URLs you used in Jekyll (see the permalink option in your _config.yml).

If there are static assets like images, they can be moved to the static/ directory in Hugo.

Then you need to use your favorite tool with some string manipulation techniques to process all Markdown files. If you use R, you can list all Markdown files and process them one by one in a loop. Below is a sketch of the code:

```
files = list.files(
  'content/', '[.](md|markdown)$', full.names = TRUE,
  recursive = TRUE
)
for (f in files) {
  blogdown:::process_file(f, function(x) {
    # process x here and return the modified x
    x
  })
}
```

The process_file() function is an internal helper function in **blogdown**. It takes a filename and a processor function to manipulate the content of the file, and writes the modified text back to the file.

To give you an idea of what a processor function may look like, I provided a few simple helper functions in **blogdown**, and below are two of them:

```
blogdown:::remove_extra_empty_lines
```

```
function (f)
process_file(f, function(x) {
    x = paste(gsub("\\s+$", "", x), collapse = "\n")
    trim_ws(gsub("\n{3,}", "\n\n", x))
})
<environment: namespace:blogdown>
```

```
blogdown:::process_bare_urls
```

```
function (f)
process_file(f, function(x) {
    gsub("\\[([^]]+)]\\(\\1/?\\)", "<\\1>", x)
})
<environment: namespace:blogdown>
```

The first function substitutes two or more empty lines with a single empty
line. The second function replaces links of the form [url](url) with
<url>. There is nothing wrong with excessive empty lines or the syntax
[url](url), though. These helper functions may make your Markdown text
a little cleaner. You can find all such helper functions at https://github.
com/rstudio/blogdown/blob/master/R/clean.R. Note they are not exported
from **blogdown**, so you need triple colons to access them.

The YAML metadata of your posts may not be completely clean, especially
when your Jekyll website was actually converted from an earlier WordPress
website. The internal helper function blogdown:::modify_yaml() may help
you clean up the metadata. For example, below is the YAML metadata of a
blog post of Simply Statistics when it was built on Jekyll:

```
---
id: 4155
title: Announcing the JHU Data Science Hackathon 2015
date: 2015-07-28T13:31:04+00:00
author: Roger Peng
layout: post
guid: http://simplystatistics.org/?p=4155
permalink: /2015/07/28/announcing-the-jhu-data-science-hackathon-2015
pe_theme_meta:
```

```
  - '0:8:"stdClass":2:{s:7:"gallery";0:8:"stdClass":...}'
al2fb_facebook_link_id:
  - 136171103105421_837886222933902
al2fb_facebook_link_time:
  - 2015-07-28T17:31:11+00:00
al2fb_facebook_link_picture:
  - post=http://simplystatistics.org/?al2fb_image=1
dsq_thread_id:
  - 3980278933
categories:
  - Uncategorized
---
```

You can discard the YAML fields that are not useful in Hugo. For example, you may only keep the fields title, author, date, categories, and tags, and discard other fields. Actually, you may also want to add a slug field that takes the base filename of the post (with the leading date removed). For example, when the post filename is 2015-07-28-announcing-the-jhu-data-science-hackathon-2015.md, you may want to add slug: announcing-the-jhu-data-science-hackathon-2015 to make sure the URL of the post on the new site remains the same.

Here is the code to process the YAML metadata of all posts:

```
for (f in files) {
  blogdown:::modify_yaml(f, slug = function(old, yaml) {
    # YYYY-mm-dd-name.md -> name
    gsub('^\\d{4}-\\d{2}-\\d{2}-|[.](md|markdown)', '', f)
  }, categories = function(old, yaml) {
    # remove the Uncategorized category
    setdiff(old, 'Uncategorized')
  }, .keep_fields = c(
    'title', 'author', 'date', 'categories', 'tags', 'slug'
  ), .keep_empty = FALSE)
}
```

You can pass a file path to modify_yaml(), define new YAML values (or func-

tions to return new values based on the old values), and decide which fields
to preserve (`.keep_fields`). You may discard empty fields via `.keep_empty = `
`FALSE`. The processed YAML metadata is below, which looks much cleaner:

```
---
title: Announcing the JHU Data Science Hackathon 2015
author: Roger Peng
date: '2015-07-28T13:31:04+00:00'
slug: announcing-the-jhu-data-science-hackathon-2015
---
```

4.2   From WordPress

From our experience, the best way to import WordPress blog posts to
Hugo is to import them to Jekyll, and write an R script to clean up the
YAML metadata of all pages if necessary, instead of using the migra-
tion tools listed on the official guide,[2] including the WordPress plugin
`wordpress-to-hugo-exporter`.

To our knowledge, the best tool to convert a WordPress website to Jekyll is
the Python tool Exitwp.[3] Its author has provided detailed instructions on
how to use it. You have to know how to install Python libraries and execute Python scripts. If you do not, I have provided an online tool at `https:`
`//github.com/yihui/travis-exitwp`. You can upload your WordPress XML
file there, and get a download link to a ZIP archive that contains your posts
in Markdown.

The biggest challenge in converting WordPress posts to Hugo is to clean
up the post content in Markdown. Fortunately, I have done this for three
different WordPress blogs,[4] and I think I have managed to automate this

---

[2] `https://gohugo.io/tools/`

[3] `https://github.com/thomasf/exitwp`

[4] The RViews blog (`https://rviews.rstudio.com`), the RStudio blog (`https://blog.`
`rstudio.com`), and Karl Broman's blog (`http://kbroman.org`). The RViews blog took me a
few days. The RStudio blog took me one day. Karl Broman's blog took me an hour.

process as much as possible. You may refer to the pull request I submitted to Karl Broman to convert his WordPress posts to Markdown (`https://github.com/kbroman/oldblog_xml/pull/1`), in which I provided both the R script and the Markdown files. I recommend that you go to the "Commits" tab and view all my GIT commits one by one to see the full process.

The key is the R script `https://github.com/yihui/oldblog_xml/blob/master/convert.R`, which converts the WordPress XML file to Markdown posts and cleans them. Before you run this script on your XML file, you have to adjust a few parameters, such as the XML filename, your old WordPress site's URL, and your new blog's URL.

Note that this script depends on the Exitwp tool. If you do not know how to run Exitwp, please use the online tool I mentioned before (travis-exitwp), and skip the R code that calls Exitwp.

The Markdown posts should be fairly clean after the conversion, but there may be remaining HTML tags in your posts, such as `<table>` and `<blockquote>`. You will need to manually clean them, if any exist.

4.3 From other systems

If you have a website built by other applications or systems, your best way to go may be to import your website to WordPress first, export it to Jekyll, and clean up the Markdown files. You can try to search for solutions like "how to import blogger.com to WordPress" or "how to import Tumblr to WordPress."

If you are very familiar with web scraping techniques, you can also scrape the HTML pages of your website, and convert them to Markdown via Pandoc, e.g.,

```
rmarkdown::pandoc_convert(
  'foo.html', to = 'markdown', output = 'foo.md'
)
```

I have actually tried this way on a website, but was not satisfied, since I still had to heavily clean up the Markdown files. If your website is simpler, this approach may work better for you.

5

Other Generators

We mentioned the possibility to bypass Hugo and use your own building method in Section D.9. Basically you have to build the site using `blogdown::build_site(method = "custom")`, and provide your own building script `/R/build.R`. In this chapter, we show you how to work with other popular static site generators like Jekyll and Hexo. Besides these static site generators written in other languages, there is actually a simple site generator written in R provided in the **rmarkdown** package (Allaire et al., 2017), and we will introduce it in Section 5.3.

5.1 Jekyll

For Jekyll (`https://jekyllrb.com`) users, I have prepared a minimal example in the GitHub repository yihui/blogdown-jekyll.[1] If you clone or download this repository and open `blogdown-jekyll.Rproj` in RStudio, you can still use all addins mentioned in Section 1.3, such as "New Post," "Serve Site," and "Update Metadata," but it is Jekyll instead of Hugo that builds the website behind the scenes now.

I assume you are familiar with Jekyll, and I'm not going to introduce the basics of Jekyll in this section. For example, you should know what the `_posts/` and `_site/` directories mean.

The key pieces of this **blogdown-jekyll** project are the files `.Rprofile`, `R/build.R`, and `R/build_one.R`. I have set some global R options for this project in `.Rprofile`:[2]

---

[1]`https://github.com/yihui/blogdown-jekyll`
[2]If you are not familiar with this file, please read Section 1.4.

```
options(
  blogdown.generator = "jekyll",
  blogdown.method = "custom",
  blogdown.subdir = "_posts"
)
```

First, the website generator was set to `jekyll` using the option `blogdown.generator`, so that **blogdown** knows that it should use Jekyll to build the site. Second, the build method `blogdown.method` was set to `custom`, so that we can define our custom R script `R/build.R` to build the Rmd files (I will explain the reason later). Third, the default subdirectory for new posts was set to `_posts`, which is Jekyll's convention. After you set this option, the "New Post" addin will create new posts under the `_posts/` directory.

When the option `blogdown.method` is `custom`, **blogdown** will call the R script `R/build.R` to build the site. You have full freedom to do whatever you want in this script. Below is an example script:

```
build_one = function(io) {
  # if output is not older than input, skip the
  # compilation
  if (!blogdown:::require_rebuild(io[2], io[1]))
    return()

  message("* knitting ", io[1])
  if (blogdown:::Rscript(shQuote(c("R/build_one.R", io))) !=
    0) {
    unlink(io[2])
    stop("Failed to compile ", io[1], " to ", io[2])
  }
}

# Rmd files under the root directory
rmds = list.files(".", "[.]Rmd$", recursive = T, full.names = T)
files = cbind(rmds, blogdown:::with_ext(rmds, ".md"))
```

```
for (i in seq_len(nrow(files))) build_one(files[i, ])

system2("jekyll", "build")
```

- Basically it contains a function `build_one()` that takes an argument `io`, which is a character vector of length 2. The first element is the input (Rmd) filename, and the second element is the output filename.

- Then we search for all Rmd files under the current directory, prepare the output filenames by substituting the Rmd file extensions with `.md`, and build the Rmd files one by one. Note there is a caching mechanism in `build_one()` that makes use of an internal **blogdown** function `require_rebuild()`. This function returns FALSE if the output file is not older than the input file in terms of the modification time. This can save you some time because those Rmd files that have been compiled before will not be compiled again every time. The key step in `build_one()` is to run the R script `R/build_one.R`, which we will explain later.

- Lastly, we build the website through a system call of the command `jekyll build`.

The script `R/build_one.R` looks like this (I have omitted some non-essential settings for simplicity):

```
local({
  # fall back on "/" if baseurl is not specified
  baseurl = blogdown:::get_config2("baseurl", default = "/")
  knitr::opts_knit$set(base.url = baseurl)
  knitr::render_jekyll()   # set output hooks

  # input/output filenames as two arguments to Rscript
  a = commandArgs(TRUE)
  d = gsub("^_|[.][a-zA-Z]+$", "", a[1])
  knitr::opts_chunk$set(
    fig.path   = sprintf("figure/%s/", d),
    cache.path = sprintf("cache/%s/", d)
  )
```

```
knitr::knit(
  a[1], a[2], quiet = TRUE, encoding = "UTF-8",
  envir = globalenv()
)
})
```

- The script is wrapped in `local()` so that an Rmd file is knitted in a clean global environment, and the variables such as `baseurl`, `a`, and `d` will not be created in the global environment, i.e., `globalenv()` used by `knitr::knit()` below.

- The **knitr** package option `base.url` is a URL to be prepended to figure paths. We need to set this option to make sure figures generated from R code chunks can be found when they are displayed on a web page. A normal figure path is often like `figure/foo.png`, and it may not work when the image is rendered to an HTML file, because `figure/foo.png` is a relative path, and there is no guarantee that this image file will be copied to the directory of the final HTML file. For example, for an Rmd source file `_posts/2015-07-23-hello.Rmd` that generates `figure/foo.png` (under `_posts/`), the final HTML file may be `_site/2015/07/23/hello/index.html`. Jekyll knows how to render an HTML file to this location, but it does not understand the image dependency and will not copy the image file to this location. To solve this issue, we render figures at the root directory `/figure/`, which will be copied to `_site/` by Jekyll. To refer to an image under `_site/figure/`, we need the leading slash (`baseurl`), e.g., `<img src="/figure/foo.png">`. This is an absolute path, so no matter where the HTML is rendered, this path always works.

- What `knitr::render_jekyll()` does is mainly to set up some **knitr** output hooks so that source code and text output from R code chunks will be wrapped in Liquid tags `{% highlight %}` and `{% end highlight %}`.

- Remember in `build.R`, we passed the variable `io` to the Rscript call `blogdown:::Rscript`. Here in `build_one.R`, we can receive them from `commandArgs(TRUE)`. The variable `a` contains an `.Rmd` and an `.md` file path. We removed the possible leading underscore (`^_`) and the extension (`[.][a-zA-Z]$`) in the path. Next we set figure and cache paths using this string. For example, for a post `_posts/foo.Rmd`, its figures will be written to figure/foo/ and its cache databases (if there are any) will be stored under cache/foo/. Both directories are under the root directory of the project.

- Lastly, we call knitr::knit() to knit the Rmd file to a Markdown output file, which will be processed by Jekyll later.

A small caveat is that since we have both .Rmd and .md files, Jekyll will treat both types of files as Markdown files by default. You have to ask Jekyll to ignore .Rmd files and only build .md files. You can set the option exclude in _config.yml:

```
exclude: ['*.Rmd']
```

Compared to the Hugo support in **blogdown**, this approach is limited in a few aspects:

1. It does not support Pandoc, so you cannot use Pandoc's Markdown. Since it uses the **knitr** package instead of **rmarkdown**, you cannot use any of **bookdown**'s Markdown features, either. You are at the mercy of the Markdown renderers supported by Jekyll.

2. Without **rmarkdown**, you cannot use HTML widgets. Basically, all you can have are dynamic text output and R graphics output from R code chunks. They may or may not suffice, depending on your specific use cases.

It may be possible for us to remove these limitations in a future version of **blogdown**, if there are enough happy Jekyll users in the R community.

5.2 Hexo

The ideas of using Hexo (https://hexo.io) are very similar to what we have applied to Jekyll in the previous section. I have also prepared a minimal example in the GitHub repository yihui/blogdown-hexo.[3]

---

[3]https://github.com/yihui/blogdown-hexo

The key components of this repository are still .Rprofile, R/build.R, and R/build_one.R. We set the option blogdown.generator to hexo, the build.method to custom, and the default subdirectory for new posts to source/_posts.

```
options(
  blogdown.generator = 'hexo',
  blogdown.method = 'custom',
  blogdown.subdir = 'source/_posts'
)
```

The script R/build.R is similar to the one in the blogdown-jekyll repository. The main differences are:

1. We find all Rmd files under the source/ directory instead of the root directory, because Hexo's convention is to put all source files under source/.

2. We call system2('hexo', 'generate') to build the website.

For the script R/build_one.R, the major difference with the script in the blogdown-jekyll repository is that we set the base.dir option for **knitr**, so that all R figures are generated to the source/ directory. This is because Hexo copies everything under source/ to public/, whereas Jekyll copies everything under the root directory to _site/.

```
local({
  # fall back on '/' if baseurl is not specified
  baseurl = blogdown:::get_config2('root', '/')
  knitr::opts_knit$set(
    base.url = baseurl, base.dir = normalizePath('source')
  )

  # input/output filenames as two arguments to Rscript
  a = commandArgs(TRUE)
  d = gsub('^source/_?|[.][a-zA-Z]+$', '', a[1])
  knitr::opts_chunk$set(
```

```
    fig.path   = sprintf('figure/%s/', d),
    cache.path = sprintf('cache/%s/', d)
  )
  knitr::knit(
    a[1], a[2], quiet = TRUE, encoding = 'UTF-8', envir = .GlobalEnv
  )
})
```

This repository is also automatically built and deployed through Netlify when I push changes to it. Since Hexo is a Node package, and Netlify supports Node, you can easily install Hexo on Netlify. For example, this example repository uses the command npm install && hexo generate to build the website; npm install will install the Node packages specified in packages.json (a file under the root directory of the repository), and hexo generate is the command to build the website from source/ to public/.

5.3 Default site generator in rmarkdown

Before **blogdown** was invented, there was actually a relatively simple way to render websites using **rmarkdown**. The structure of the website has to be a flat directory of Rmd files (no subdirectories for Rmd files) and a configuration file in which you can specify a navigation bar for all your pages and output format options.

You can find more information about this site generator in its documentation at http://rmarkdown.rstudio.com/rmarkdown_websites.html, and we are not going to repeat the documentation here, but just want to highlight the major differences between the default site generator in **rmarkdown** and other specialized site generators like Hugo:

- The **rmarkdown** site generator requires all Rmd files to be under the root directory. Hugo has no constraints on the site structure, and you can create arbitrary directories and files under /content/.

- Hugo is a general-purpose site generator that is highly customizable, and there are a lot of things that **rmarkdown**'s default site generator does not support, e.g., RSS feeds, metadata especially common in blogs such as categories and tags, and customizing permanent links for certain pages.

There are still legitimate reasons to choose the **rmarkdown** default site generator, even though it does not appear to be as powerful as Hugo, including:

- You are familiar with generating single-page HTML output from R Markdown, and all you want is to extend this to generating multiple pages from multiple Rmd files.

- It suffices to use a flat directory of Rmd files. You do not write a blog or need RSS feeds.

- You prefer the Bootstrap styles. In theory, you can also apply Bootstrap styles to Hugo websites, but it will require you to learn more about Hugo. Bootstrap is well supported in **rmarkdown**, and you can spend more time on the configurations instead of learning the technical details about how it works.

- There are certain features in **rmarkdown** HTML output that are missing in **blogdown**. For example, currently you cannot easily print data frames as paged tables, add a floating table of contents, or fold/unfold code blocks dynamically in the output of **blogdown**. All these could be implemented via JavaScript and CSS, but it is certainly not as simple as specifying a few options in **rmarkdown** like `toc_float: true`.

Please note that the **rmarkdown** site generator is extensible, too. For example, the **bookdown** package (Xie, 2017b) is essentially a custom site generator to generate books as websites.

5.4   pkgdown

The **pkgdown** package (Wickham (2017), `https://github.com/hadley/ pkgdown`) can help you quickly turn the R documentation of an R package (including help pages and vignettes) into a website. It is independent of **blogdown** and solves a specific problem. It is not a general-purpose website generator. We want to mention it in this book because it is very easy to use, and also highly useful. You can find the instructions on its website or in its GitHub repository.

A

R Markdown

R Markdown (Allaire et al., 2017) is a plain-text document format consisting of two components: R (or other computing languages) and Markdown. Markdown makes it easy for authors to write a document due to its simple syntax. Program code (such as R code) can be embedded in a source Markdown document to generate an output document directly: when compiling the source document, the program code will be executed and its output will be intermingled with the Markdown text.

R Markdown files usually use the filename extension .Rmd. Below is a minimal example:

```
---
title: A Simple Linear Regression
author: Yihui Xie
---

We build a linear regression below.

```{r}
fit = lm(dist ~ speed, data = cars)
b = coef(summary(fit))
plot(fit)
```

The slope of the regression is `r b[2, 1]`.
```

Such a document can be compiled using the function rmarkdown::render(), or equivalently, by clicking the Knit button in RStudio. Under the hood, an R Markdown document is first compiled to Markdown through **knitr** (Xie, 2017c), which executes all program code in the document. Then the Markdown output document is compiled to the final output document through Pandoc, such as an HTML page, a PDF document, a Word document, and so on. It is important to know this two-step process, otherwise you may not know which package documentation to look up when you have questions. Basically, for anything related to the (R) code chunks, consult the **knitr** documentation (`https://yihui.name/knitr/`); for anything related to Markdown, consult the Pandoc documentation (`https://pandoc.org`).

An R Markdown document typically consists of YAML metadata (optional) and the document body. YAML metadata are written between a pair of `---` to set some attributes of the document, such as the title, author, and date, etc. In the document body, you can mix code chunks and narratives. A code block starts with a chunk header ```` ```{r} ```` and ends with ```` ``` ````. There are many possible chunk options that you can set in the chunk header to control the output, e.g., you can set the figure height to 4 inches using ```` ```{r fig.height=4} ````. For all possible chunk options, see `https://yihui.name/knitr/options/`.

Pandoc supports a large variety of output document formats. For **blogdown**, the output format is set to HTML (`blogdown::html_page`), since a website typically consists of HTML pages. If you want other formats, please see Section 2.7. To create an R Markdown post for **blogdown**, it is recommended that you use the RStudio "New Post" (Figure 1.2) or the function `blogdown::new_post()`, instead of the RStudio menu `File -> New File -> R Markdown`.

You are strongly recommended to go through the documentation of **knitr** chunk options and Pandoc's manual at least once to have an idea of all possibilities. The basics of Markdown are simple enough, but there are many less well-known features in Pandoc's Markdown, too. As we mentioned in Section 1.5, **blogdown**'s output format is based on **bookdown** (Xie, 2017b), which contains several other Markdown extensions, such as numbered equations and theorem environments, and you need to read Chapter 2 of the **bookdown** book (Xie, 2016) to learn more about these features.

You can find an R Markdown cheat sheet and a reference guide at `https://www.rstudio.com/resources/cheatsheets/`, which can be handy after you are more familiar with R Markdown.

With R Markdown, you only need to maintain the source documents; all output pages can be automatically generated from source documents. This makes it much easier to maintain a website, especially when the website is related to data analysis or statistical computing and graphics. When the source code is updated (e.g., the model or data is changed), your web pages can be updated accordingly and automatically. There is no need to run the code separately and cut-and-paste again. Besides the convenience, you gain reproducibility at the same time.

B

Website Basics

If you want to tweak the appearance of your website, or even design your own theme, you must have some basic knowledge of web development. In this appendix, we briefly introduce HTML, CSS, and JavaScript, which are the most common components of a web page, although CSS and JavaScript are optional.

We only aim at getting you started with HTML, CSS, and JavaScript. HTML is relatively simple to learn, but CSS and JavaScript can be much more complicated, depending on how much you want to learn and what you want to do with them. After reading this appendix, you will have to use other resources to teach yourself. When you search for these technologies online, the most likely websites that you may hit are MDN[1] (Mozilla Developer Network), w3schools.com,[2] and StackOverflow.[3] Among these websites, w3schools often provides simple examples and tutorials that may be friendlier to beginners, but we often hear negative comments[4] about it, so please use it with caution. I often read all three websites when looking for solutions.

If we were only allowed to give one single most useful tip about web development, it would be: use the Developer Tools of your web browser. Most modern web browsers provide these tools. For example, you can find these tools from the menu of Google Chrome `View -> Developer`, Firefox's menu `Tools -> Web Developer`, or Safari's menu `Develop -> Show Web Inspector`. Figure B.1 is a screenshot of using the Developer Tools in Chrome.

Typically you can also open the Developer Tools by right-clicking on a certain element on the web page and selecting the menu item `Inspect` (or `Inspect Element`). In Figure B.1, I right-clicked on the profile image of my website

---

[1] `https://developer.mozilla.org`
[2] `https://www.w3schools.com`
[3] `https://stackoverflow.com`
[4] `https://meta.stackoverflow.com/q/280478/559676`

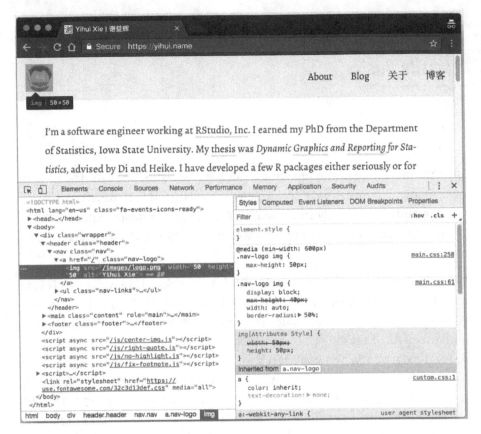

FIGURE B.1: Developer Tools in Google Chrome.

`https://yihui.name` and inspected it, and Chrome highlighted its HTML source code `<img src="..." ... />` in the left pane. You can also see the CSS styles associated with this img element in the right pane. What is more, you can interactively change the styles there if you know CSS, and see the (temporary) effects in real time on the page! After you are satisfied with the new styles, you can write the CSS code in files.

There are a lot of amazing features of Developer Tools, which make them not only extremely useful for debugging and experimentation, but also helpful for learning web development. These tools are indispensable to me when I develop anything related to web pages. I learned a great deal about CSS and JavaScript by playing with these tools.

B.1   HTML

HTML stands for Hyper Text Markup Language, and it is the language behind most web pages you see. You can use the menu View -> View Source
or the context menu View Page Source to see the full HTML source of a web
page in your browser. All elements on a page are represented by HTML tags.
For example, the tag <p> represents paragraphs, and <img> represents images.

The good thing about HTML is that the language has only a limited number
of tags, and the number is not very big (especially the number of commonly
used tags). This means there is hope that you can master this language fully
and quickly.

Most HTML tags appear in pairs, with an opening tag and a closing tag,
e.g., <p></p>. You write the content between the opening and closing tags,
e.g., <p>This is a paragraph.</p>. There are a few exceptions, such as the
<img> tag, which can be closed by a slash / in the opening tag, e.g., <img
src="foo.png" />. You can specify attributes of an element in the opening
tag using the syntax name=value (a few attributes do not require value).

HTML documents often have the filename extension .html (or .htm). Below
is an overall structure of an HTML document:

```
<html>

  <head>
  </head>

  <body>
  </body>

</html>
```

Basically an HTML document consists a head section and body section. You
can specify the metadata and include assets like CSS files in the head section.

Normally the head section is not visible on a web page. It is the body section that holds the content to be displayed to a reader. Below is a slightly richer example document:

```html
<!DOCTYPE html>
<html>

  <head>
    <meta charset="utf-8" />

    <title>Your Page Title</title>

    <link rel="stylesheet" href="/css/style.css" />
  </head>

  <body>
    <h1>A First-level Heading</h1>

    <p>A paragraph.</p>

    <img src="/images/foo.png" alt="A nice image" />

    <ul>
      <li>An item.</li>
      <li>Another item.</li>
      <li>Yet another item.</li>
    </ul>

    <script src="/js/bar.js"></script>
  </body>

</html>
```

In the head, we declare the character encoding of this page to be UTF-8 via a <meta> tag, specify the title via the <title> tag, and include a stylesheet via a <link> tag.

The body contains a first-level section heading `<h1>`,[5] a paragraph `<p>`, an image `<img>`, an unordered list `<ul>` with three list items `<li>`, and includes a JavaScript file in the end via `<script>`.

There are much better tutorials on HTML than this section, such as those offered by MDN and w3schools.com, so we are not going to make this section a full tutorial. Instead, we just want to provide a few tips on HTML:

- You may validate your HTML code via this service: `https://validator.w3.org`. This validator will point out potential problems of your HTML code. It actually works for XML and SVG documents, too.

- Among all HTML attributes, file paths (the `src` attribute of some tags like `<img>`) and links (the `href` attribute of the `<a>` tag) may be the most confusing to beginners. Paths and links can be relative or absolute, and may come with or without the protocol and domain. You have to understand what a link or path exactly points to. A full link is of the form `http://www.example.com/foo/bar.ext`, where `http` specifies the protocol (it can be other protocols like `https` or `ftp`), `www.example.com` is the domain, and `foo/bar.ext` is the file under the root directory of the website.

  - If you refer to resources on the same website (the same protocal and domain), we recommend that you omit the protocol and domain names, so that the links will continue to work even if you change the protocol or domain. For example, a link `<a href="/hi/there.html">` on a page `http://example.com/foo/` refers to `http://example.com/hi/there.html`. It does not matter if you change `http` to `https`, or `example.com` to `another-domain.com`.

  - Within the same website, a link or path can be relative or absolute. The meaning of an absolute path does not change no matter where the current HTML file is placed, but the meaning of a relative path depends on the location of the current HTML file. Suppose you are currently viewing the page `example.com/hi/there.html`:

    * A absolute path `/foo/bar.ext` always means `example.com/foo/bar.ext`. The leading slash means the root directory of the website.

---

[5]There are six possible levels from h1, h2, ..., to h6.

* A relative path `../images/foo.png` means `example.com/images/foo.png` (`..` means to go one level up). However, if the HTML file `there.html` is moved to `example.com/hey/again/there.html`, this path in `there.html` will refer to `example.com/hey/images/foo.png`.

* When deciding whether to use relative or absolute paths, here is the rule of thumb: if you will not move the resources referred or linked to from one subpath to another (e.g., from `example.com/foo/` to `example.com/bar/`), but only move the HTML pages that use these resources, use absolute paths; if you want to change the subpath of the URL of your website, but the relative locations of HTML files and the resources they use do not change, you may use relative links (e.g., you can move the whole website from `example.com/` to `example.com/foo/`).

* If the above concepts sound too complicated, a better way is to either think ahead carefully about the structure of your website and avoid moving files, or use rules of redirects if supported (such as 301 or 302 redirects).

– If you link to a different website or web page, you have to include the domain in the link, but it may not be necessary to include the protocol, e.g., `//test.example.com/foo.css` is a valid path. The actual protocol of this path matches the protocol of the current page, e.g., if the current page is `https://example.com/`, this link means `https://test.example.com/foo.css`. It may be beneficial to omit the protocol because HTTP resources cannot be embedded on pages served through HTTPS (for security reasons), e.g., an image at `http://example.com/foo.png` cannot be embedded on a page `https://example.com/hi.html` via `<img src="http://example.com/foo.png" />`, but `<img src="//example.com/foo.png" />` will work if the image can be accessed via HTTPS, i.e., `https://example.com/foo.png`. The main drawback of not including the protocol is that such links and paths do not work if you open the HTML file locally without using a web server, e.g., only double-click the HTML file in your file browser and show it in the browser.[6]

– A very common mistake that people make is a link without the leading double slashes before the domain. You may think `www.example.com` is a valid link. It is not! At least it does not link to the website that you intend to link to. It works when you type it in the address bar of your browser because your browser will normally autocomplete it to `http://www.example.com`. However, if you write a link `<a href="www.example.com">See this link</a>`, you will be in trouble. The browser will interpret this as a relative link, and it is relative to the URL of the current web page, e.g., if you are currently viewing `http://yihui.name/cn/`, the link `www.example.com` actually means `http://yihui.name/cn/www.example.com`! Now you should know the Markdown text `[Link](www.example.com)` is typically a mistake, unless you really mean to link to a subdirectory of the current page or a file with literally the name `www.example.com`.

B.2 CSS

The Cascading Stylesheets (CSS) language is used to describe the look and formatting of documents written in HTML. CSS is responsible for the visual style of your site. CSS is a lot of fun to play with, but it can also easily steal your time.

In the Hugo framework (`https://gohugo.io/tutorials/creating-a-new-theme/`), CSS is one of the major "non-content" files that shapes the look and feel of your site (the others are images, JavaScript, and Hugo templates). What does the "look and feel"[7] of a site mean? "Look" generally refers to static style components including, but not limited to

---

[6]That is because without a web server, an HTML file is viewed via the protocol `file`. For example, you may see a URL of the form `file://path/to/the/file.html` in the address bar of your browser. The path `//example.com/foo.png` will be interpreted as `file://example.com/foo.png`, which is unlikely to exist as a local file on your computer.

[7]`https://en.wikipedia.org/wiki/Look_and_feel`

- color palettes,
- images,
- layouts/margins, and
- fonts.

whereas "feel" relates to dynamic components that the user interacts with like

- dropdown menus,
- buttons, and
- forms.

There are 3 ways to define styles. The first is in line with HTML. For example, this code

```
<p>Marco! <em>Polo!</em></p>
```

would produce text that looks like this: Marco! *Polo!*

However, this method is generally not preferred for numerous reasons.[8]

A second way is to internally define the CSS by placing a style section in your HTML:

```
<html>
<style>
#favorite {
    font-style: italic;
}
</style>
<ul id="tea-list">
  <li>Earl Grey</li>
  <li>Darjeeling</li>
  <li>Oolong</li>
  <li>Chamomile</li>
  <li id="favorite">Chai</li>
```

---

[8] https://stackoverflow.com/q/12013532/559676

```
    </ul>
</html>
```

In this example, only the last tea listed, *Chai*, is italicized.

The third method is the most popular because it is more flexible and the least repetitive. In this method, you define the CSS in an external file that is then referenced as a link in your HTML:

```
<html>
    <link rel="stylesheet" href="/css/style.css" />
</html>
```

What goes inside the linked CSS document is essentially a list of rules (the same list could appear internally between the style tags, if you are using the second method). Each rule must include both a selector or group of selectors, and a declarations block within curly braces that contains one or more property: value; pairs. Here is the general structure for a rule[9]:

```
/* CSS pseudo-code */
selectorlist {
    property: value;
    /* more property: value; pairs*/
}
```

Selectors[10] can be based on HTML element types or attributes, such as id or class (and combinations of these attributes):

```
/* by element type */
li {
    color: yellow; /* all <li> elements are yellow */
}


/* by ID with the # symbol */
```

---

[9]https://developer.mozilla.org/en-US/docs/Web/CSS/Reference
[10]https://developer.mozilla.org/en-US/docs/Web/CSS/Reference#Selectors

```
#my-id {
    color: yellow; /* elements with id = "my-id" are yellow */
}


/* by class with the . symbol */
.my-class {
    color: yellow; /* elements with class = "my-class" are yellow */
}
```

Because each HTML element may match several different selectors, the CSS standard determines which set of rules has precedence for any given element, and which properties to inherit. This is where the cascade algorithm comes into play. For example, take a simple unordered list:

```
<ul id="tea-list">
  <li>Earl Grey</li>
  <li>Darjeeling</li>
  <li>Oolong</li>
  <li>Chamomile</li>
  <li>Chai</li>
</ul>
```

Now, let's say we want to highlight our favorite teas again, so we'll use a class attribute.

```
<ul id="tea-list">
  <li>Earl Grey</li>
  <li class="favorite">Darjeeling</li>
  <li>Oolong</li>
  <li>Chamomile</li>
  <li class="favorite">Chai</li>
</ul>
```

We can use this class attribute as a selector in our CSS. Let's say we want our favorite teas to be in bold and have a background color of yellow, so our CSS would look like this:

```
.favorite {
  font-weight: bold;
  background-color: yellow;
}
```

Now, if you want every list item to be italicized with a white background, you can set up another rule:

```
li {
  font-style: italic;
  background-color: white;
}
```

If you play with this code (which you can do easily using sites like http://jsbin.com, https://jsfiddle.net, or https://codepen.io/pen/), you'll see that the two favorite teas are still highlighted in yellow. This is because the CSS rule about .favorite as a class is more specific than the rule about li type elements. To override the .favorite rule, you need to be as specific as you can be when choosing your group of selectors:

```
ul#tea-list li.favorite {
  background-color: white;
}
```

This example just scratches the surface of cascade and inheritance.[11]

For any Hugo theme that you install, you can find the CSS file in the themes/ folder. For example, the default lithium theme is located in: themes/hugo-lithium-theme/static/css/main.css. Once you are familiar with CSS, you can understand how each set of rules work to shape the visual style of your website, and how to alter the rules. For some themes (i.e., the hugo-academic theme[12]), you have the option of linking to a custom CSS,[13] which you can use to further customize the visual style of your site.

---

[11]https://developer.mozilla.org/en-US/docs/Learn/CSS/Introduction_to_CSS/Cascade_and_inheritance

[12]https://github.com/gcushen/hugo-academic

[13]https://gist.github.com/gcushen/d5525a4506b9ccf83f2bce592a895495

A few one-line examples illustrate how simple CSS rules can be used to make dramatic changes:

- To make circular or rounded images, you may assign a class `img-circle` to images (e.g., `<img class="img-circle" src="foo.png" />`) and define the CSS:

```
.img-circle {
  border-radius: 50%;
}
```

- To make striped tables, you can add background colors to odd or even rows of the table, e.g.,

```
tr:nth-child(even) {
  background: #eee;
}
```

- You can append or prepend content to elements via pseudo-elements `::after` and `::before`. Here is an example of adding a period after section numbers: `https://github.com/rstudio/blogdown/issues/80`.

B.3 JavaScript

It is way more challenging to briefly introduce JavaScript than HTML and CSS, since it is a programming language. There are many books and tutorials about this language. Anyway, we will try to scratch the surface for R users in this section.

In a nutshell, JavaScript is a language that is typically used to manipulate elements on a web page. An effective way to learn it is through the JavaScript console in the Developer Tools of your web browser (see Figure B.1), because you can interactively type code in the console and execute it, which feels similar to executing R code in the R console (e.g., in RStudio). You may open

any web page in your web browser (e.g., https://yihui.name), then open the
JavaScript console, and try the code below on any web page:

```
document.body.style.background = 'orange';
```

It should change the background color of the page to orange, unless the page
has already defined background colors for certain elements.

To effectively use JavaScript, you have to learn both the basic syntax of
JavaScript and how to select elements on a page before you can manipulate
them. You may partially learn the former from the short JavaScript snippet
below:

```
var x = 1;   // assignments
1 + 2 - 3 * 4 / 5;   // arithmetic


if (x < 2) console.log(x);   // "print" x


var y = [9, 1, 0, 2, 1, 4];   // array

// function
var sum = function(x) {
  var s = 0;
  // a naive way to compute the sum
  for (var i=0; i < x.length; i++) {
    s += x[i];
  }
  return s;
};


sum(y);


var y = "Hello World";
y = y.replace(" ", ", ");   // string manipulation
```

You may feel the syntax is similar to R to some degree. JavaScript is an object-
oriented language, and usually there are several methods that you can ap-

ply to an object. The string manipulation above is a typical example of the Object.method() syntax. To know the possible methods on an object, you can type the object name in your JavaScript console followed by a dot, and you should see all candidates.

R users have to be extremely cautious because JavaScript objects are often mutable, meaning that an object could be modified anywhere. Below is a quick example:

```
var x = {"a": 1, "b": 2};  // like a list in R
var f = function(z) {
  z.a = 100;
};
f(x);
x;  // modified! x.a is 100 now
```

There are many mature JavaScript libraries that can help you select and manipulate elements on a page, and the most popular one may be jQuery.[14] However, you should know that sometimes you can probably do well enough without these third-party libraries. There are some basic methods to select elements, such as document.getElementById() and document.getElementsByClassName(). For example, you can select all paragraphs using document.querySelectorAll('p').

Next we show a slightly advanced application, in which you will see anonymous functions, selection of elements by HTML tag names, regular expressions, and manipulation of HTML elements.

In Section 2.5.2, we mentioned how to enable MathJax on a Hugo website. The easy part is to include the script MathJax.js via a <script> tag, and there are two hard parts:

1. How to protect the math content from the Markdown engine (Blackfriday), e.g., we need to make sure underscores in math expressions are not interpreted as <em></em>. This problem only exists in plain Markdown posts, and has been mentioned in Section 1.5 without explaining the solution.

---

[14]https://jquery.com

2. By default, MathJax does not recognize a pair of single dollar signs as the syntax for inline math expressions, but most users are perhaps more comfortable with the syntax x than \(x\).

The easiest solution to the first problem may be adding backticks around math expressions, e.g., `` `$x_i$` ``, but the consequence is that the math expression will be rendered in <code></code>, and MathJax ignores <code> tags when looking for math expressions on the page. We can force MathJax to search for math expressions in <code>, but this will still be problematic. For example, someone may want to display inline R code `` `list$x$y` ``, and x may be recognized as a math expression. MathJax ignores <code> for good reasons. Even if you do not have such expressions in <code>, you may have some special CSS styles attached to <code>, and these styles will be applied to your math expressions, which can be undesired (e.g., a light gray background).

To solve these problems, I have provided a solution in the JavaScript code at https://yihui.name/js/math-code.js:

```javascript
(function() {
  var i, text, code, codes = document.getElementsByTagName('code');
  for (i = 0; i < codes.length;) {
    code = codes[i];
    if (code.parentNode.tagName !== 'PRE' &&
        code.childElementCount === 0) {
      text = code.textContent;
      if (/^\$[^$]/.test(text) && /[^$]\$$/.test(text)) {
        text = text.replace(/^\$/, '\\(').replace(/\$$/, '\\)');
        code.textContent = text;
      }
      if (/^\\\((.|\s)+\\\)$/.test(text) ||
          /^\\\[(.|\s)+\\\]$/.test(text) ||
          /^\$(.|\s)+\$$/.test(text) ||
          /^\\begin\{([^}]+)\}(.|\s)+\\end\{[^}]+\}$/.test(text)) {
        code.outerHTML = code.innerHTML;  // remove <code></code>
        continue;
      }
    }
```

```
    }
    i++;
  }
})();
```

It is not a perfect solution, but it should be very rare that you run into problems. This solution identifies possible math expressions in <code>, and strips the <code> tag, e.g., replaces <code>x</code> with \(x\). After this script is executed, we load the MathJax script. This way, we do not need to force MathJax to search for math expressions in <code> tags, and your math expressions will not inherit any styles from <code>. The JavaScript code above is not too long, and should be self-explanatory. The trickiest part is i++. I will leave it to readers to figure out why the for loop is not the usual form for (i = 0; i < codes.length; i++). It took me quite a few minutes to realize my mistake when I wrote the loop in the usual form.

B.4   Useful resources

B.4.1   File optimization

Although static websites are fast in general, you can certainly further optimize them. You may search for "CSS and JavaScript minifier," and these tools can compress your CSS and JavaScript files, so that they can be loaded faster. Since there are a lot of tools, I will not recommend any here.

You can also optimize images on your website. I frequently use a command-line tool named optipng[15] to optimize PNG images. It is a lossless optimizer, meaning that it reduces the file size of a PNG image without loss of quality. From my experience, it works very well on PNG images generated from R, and can reduce the file size by at least 30% (sometimes even more than 50%). Personally I also use online tools like http://optimizilla.com to op-

---

[15] http://optipng.sourceforge.net

timize PNG and JPEG images. For GIF images, I often use `https://ezgif.com/optimize` to reduce the file sizes if they are too big.

Note that Netlify has provided the optimization features on the server side for free at the moment, so you may just want to enable them there instead of doing all the hard work by yourself.

B.4.2   Helping people find your site

Once your site is up and running, you probably want people to find it. SEO — Search Engine Optimization — is the art of making a website easy for search engines like Google to understand. And, hopefully, if the search engine knows what you are writing about, it will present links to your site high in results when someone searches for topics you cover.

Entire books have been written about SEO, not to mention the many companies that are in the business of offering (paid) technical and strategic advice to help get sites atop search-engine rankings. If you are interested in finding out more, a good place to start is the Google Search Engine Optimization Starter Guide (`http://bit.ly/google-seo-starter`). Below are a few key points:

1. The title that you select for each page and post is a very important signal to Google and other search engines telling them what that page is about.

2. Description tags are also critical to explain what a page is about. In HTML documents, description tags[16] are one way to provide metadata about the page. Using **blogdown**, the description may end up as text under the page title in a search-engine result. If your page's YAML does not include a description, you can add one like `description: "A brief description of this page.";` the HTML source of the rendered page would have a `<meta>` tag in `<head>` like `<meta name="description" content="A brief description of this page.">`. Not all themes support adding the description to your HTML page (although they should!)

---

[16] `https://www.w3schools.com/tags/tag_meta.asp`

3.   URL structure also matters. You want your post's slug to have
     informative keywords, which gives another signal of what the
     page is about. Have a post with interesting things to do in San
     Francisco? `san-francisco-events-calendar` might be a better slug
     than `my-guide-to-fun-things-to-do`.

C

Domain Name

While you can use the free subdomain names like those provided by GitHub or Netlify, it may be a better idea to own a domain name of your own. The cost of an apex domain is minimal (typically the yearly cost is about US$10), and you will enter a much richer world after you purchase a domain name. For example, you are free to point your domain to any web servers, you can create as many subdomain names as you want, and you can even set up your own email accounts using the domain or subdomains. In this appendix, we will explain some basic concepts of domain names, and mention a few (free) services to help you configure your domain name.

Before we dive into the details, we want to outline the big picture of how a URL works in your web browser. Suppose you typed or clicked a link `http://www.example.com/foo/index.html` in your web browser. What happens behind the scenes before you see the actual web page?

First, the domain name has to be resolved through the nameservers associated with it. A nameserver knows the DNS (Domain Name System) records of a domain. Typically it will look up the "A records" to point the domain to the IP address of a web server. There are several other types of DNS records, and we will explain them later. Once the web server is reached, the server will look for the file `foo/index.html` under a directory associated with the domain name, and return its content in the response. That is basically how you can see a web page.

C.1   Registration

You can purchase a domain name from many domain name registrars. To stay neutral, we are not going to make recommendations here. You can use your search engine to find a registrar by yourself, or ask your friends for recommendations. However, we would like to remind you of a few things that you should pay attention to when looking for a domain name registrar:

- You should have the freedom to transfer your domain from the current registrar to other registrars, i.e., they should not lock you in their system. To transfer a domain name, you should be given a code known as the "Transfer Auth Code" or "Auth Code" or "Transfer Key" or something like that.

- You should be able to customize the nameservers (see Section C.2) of your domain. By default, each registrar will assign their own nameservers to you, and these nameservers typically work very well. However, there are some special nameservers that provide services more than just DNS records, and you may be interested in using them.

- Other people can freely look up your personal information, such as your email or postal address, after you register a domain and submit this information to the registrar. This is called the "WHOIS Lookup." You may want to protect your privacy, but your registrar may require an extra payment.

C.2   Nameservers

The main reason why we need nameservers is that we want to use domains instead of IP addresses, although a domain is not strictly necessary for you to be able to access a website. You could use the IP address if you have your own server with a public IP, but there are many problems with this approach. For example, IP addresses are limited (in particular IPv4), not easy to memorize, and you can only host one website per IP address (without using other ports).

A nameserver is an engine that directs the DNS records of your domain. The most common DNS record is the A record, which maps a domain to an IP address, so that the hosting server can be found via its IP address when a website is accessed through a domain. We will introduce two more types of DNS records in Section C.3: CNAME and MX records.

In most cases, the default nameservers provided by your domain registrar should suffice, but there is a special technology missing in most nameservers: CNAME flattening. You only need this technology if you want to set a CNAME record for your apex domain. The only use case to my knowledge is when you host your website via Netlify, but want to use the apex domain instead of the www subdomain, e.g., you want to use example.com instead of www.example.com. To make use of this technology, you could consider Cloudflare,[1] which provides this DNS feature for free. Basically, all you need to do is to point the nameservers of your domain to the nameservers provided by Cloudflare (of the form *.ns.cloudflare.com).

C.3 DNS records

There are many types of DNS records, and you may see a full list on Wikipedia.[2] The most commonly used types may be A, CNAME, and MX records. Figure C.1 shows a subset of DNS records of my domain yihui.name on Cloudflare, which may give you an idea of what DNS records look like. You may query DNS records using command-line tools such as dig[3] or an app provided by Google: https://toolbox.googleapps.com/apps/dig/.

An apex domain can have any number of subdomains. You can set DNS records for the apex domain and any subdomains. You can see from Figure C.1 that I have several subdomains, e.g., slides.yihui.name and xran.yihui.name.

As we have mentioned, an A record points a domain or subdomain to an IP address of the host server. I did not use any A records for my domains

---

[1]https://www.cloudflare.com
[2]https://en.wikipedia.org/wiki/List_of_DNS_record_types
[3]https://en.wikipedia.org/wiki/Dig_(command)

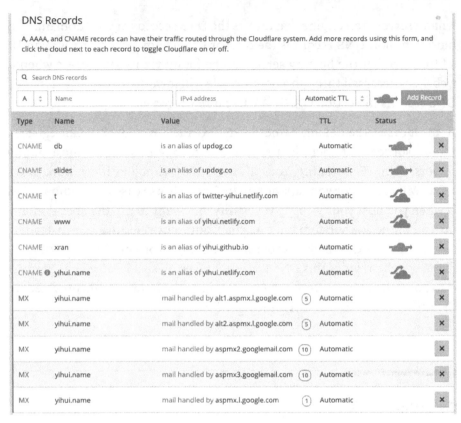

FIGURE C.1: Some DNS records of the domain yihui.name on Cloudflare.

since all services I use, such as Updog, GitHub Pages, and Netlify, support CNAME records well. A CNAME record is an alias, pointing one domain to another domain. The advantage of using CNAME over A is that you do not have to tie a domain to a fixed IP address. For example, the CNAME record for t.yihui.name is twitter-yihui.netlify.com. The latter domain is provided by Netlify, and I do not need to know where they actually host the website. They are free to move the host of twitter-yihui.netlify.com, and I will not need to update my DNS record. Every time someone visits the website t.yihui.name, the web browser will route the traffic to the domain set in the CNAME record. Note that this is different from redirection, i.e., the URL t.yihui.name will not be explicitly redirected to twitter-yihui.netlify.com (you still see the former in the address bar of your browser).

Normally, you can set any DNS records for the apex domain except CNAME, but I set a CNAME record for my apex domain y i hu i . name, and that is because Cloudflare supports CNAME flattening. For more information on this topic, you may read the post "To WWW or not WWW,"[4] by Netlify. Personally, I prefer not using the subdomain www. y i hu i . name to keep my URLs short, so I set a CNAME record for both the apex domain y i hu i . name and the www subdomain, and Netlify will automatically redirect the www subdomain to the apex domain. That said, if you are a beginner, it may be a little easier to configure and use the www subdomain, as suggested by Netlify. Note www is a conventional subdomain that sounds like an apex domain, but really is not; you can follow this convention or not as you wish.

For email services, I was an early enough "netizen",[5] and when I registered my domain name, Google was still offering free email services to custom domain owners. That is how I can have a custom mailbox x i e@yihui . name. Now you will have to pay for G Suite.[6] In Figure C.1 you can see I have set some MX (stands for "mail exchange") records that point to some Google mail servers. Of course, Google is not the only possible choice when it comes to custom mailboxes. Migadu[7] claims to be the "most affordable email hosting." You may try its free plan and see if you like it. Unless you are going to use your custom mailbox extensively and for professional purposes, the free plan may suffice. In fact, you may create an alias address on Migadu to forward emails to your other email accounts (such as Gmail) if you do not care about an actual custom mailbox. Migadu has provided detailed instructions on how to set the MX records for your domain.

---

[4]https://www.netlify.com/blog/2017/02/28/to-www-or-not-www/
[5]https://en.wikipedia.org/wiki/Netizen
[6]https://gsuite.google.com
[7]https://www.migadu.com

D

Advanced Topics

In this appendix, we talk about a few advanced topics that may be of interest to developers and advanced users.

D.1 More global options

There are a few more advanced global options in addition to those introduced in Section 1.4, and they are listed in Table D.1.

If you want to install Hugo to a custom path, you can set the global option `blogdown.hugo.dir` to a directory to store the Hugo executable before you call `install_hugo()`, e.g., `options(blogdown.hugo.dir = '~/Downloads/hugo_0.20.1/')`. This may be useful for you to use a specific version of Hugo for a specific website,[1] or store a copy of Hugo on a USB Flash drive along with your website.

The option `blogdown.method` is explained in Section D.9.

---

[1] You can set this option per project. See Section 1.4 for details.

TABLE D.1: A few more advanced global options.

Option name	Default	Meaning
blogdown.hugo.dir		The directory of the Hugo executable
blogdown.method	html	The building method for R Markdown
blogdown.publishDir		The publish dir for local preview
blogdown.widgetsID	TRUE	Incremental IDs for HTML widgets?

When your website project is under version control in the RStudio IDE, continuously previewing the site can be slow, if it contains hundreds of files or more. The default publish directory is `public/` under the project root directory, and whenever you make a change in the source that triggers a rebuild, RStudio will be busy tracking file changes in the `public/` directory. The delay before you see the website in the RStudio Viewer can be 10 seconds or even longer. That is why we provide the option `blogdown.publishDir`. You may set a temporary publish directory to generate the website, and this directory should not be under the same RStudio project, e.g., `options(blogdown.publishDir = '../public_site')`, which means the website will be generated to the directory `public_site/` under the parent directory of the current project.

The option `blogdown.widgetsID` is only relevant if your website source is under version control and you have HTML widgets on the website. If this option is `TRUE` (default), the random IDs of HTML widgets will be changed to incremental IDs in the HTML output, so these IDs are unlikely to change every time you recompile your website; otherwise, every time you will get different random IDs.

D.2   LiveReload

As we briefly mentioned in Section 1.2, you can use `blogdown::serve_site()` to preview a website, and the web page will be automatically rebuilt and reloaded in your web browser when the source file is modified and saved. This is called "LiveReload."

We have provided two approaches to LiveReload. The default approach is through `servr::httw()`, which will continuously watch the website directory for file changes, and rebuild the site when changes are detected. This approach has a few drawbacks:

1.   It is relatively slow because the website is fully regenerated every time. This may not be a real problem for Hugo, because Hugo is often fast enough: it takes about a millisecond to generate one page,

so a website with a thousand pages may only take about one second to be fully regenerated.

2.   The daemonized server (see Section 1.4) may not work.

If you are not concerned about the above issues, we recommend that you use the default approach, otherwise you can set the global option `options(blogdown.generator.server = TRUE)` to use an alternative approach to LiveReload, which is based on the native support for LiveReload from the static site generator. At the moment, this has only been tested against Hugo-based websites. It does not work with Jekyll and we were not successful with Hexo, either.

This alternative approach requires two additional R packages to be installed: **processx** (Csárdi, 2017) and **later** (Cheng, 2017). You may use this approach when you primarily work on plain Markdown posts instead of R Markdown posts, because it can be much faster to preview Markdown posts using the web server of Hugo. The web server can be stopped by `blogdown::stop_server()`, and it will always be stopped when the R session is ended, so you can restart your R session if `stop_server()` fails to stop the server for some reason.

The web server is established via the command `hugo server` (see its documentation[2] for details). You can pass command-line arguments via the global option `blogdown.hugo.server`. The default value for this option is `c('-D', '-F')`, which means to render draft and future posts in the preview. We want to highlight a special argument `--navigateToChanged` in a recent version of Hugo, which asks Hugo to automatically navigate to the changed page. For example, you can set the options:

```
options(blogdown.hugo.server = c("-D", "-F", "--navigateToChanged"))
```

Then, when you edit a source file under `content/`, Hugo will automatically show you the corresponding output page in the web browser.

Note that Hugo renders and serves the website from memory by default, so no files will be generated to `public/`. If you need to publish the

---

[2]https://gohugo.io/commands/hugo_server/

`public/` folder manually, you will have to manually build the website via `blogdown::hugo_build()` or `blogdown::build_site()`.

D.3   Building a website for local preview

The function `blogdown::build_site()` has an argument `local` that defaults to `FALSE`, which means building the website for publishing instead of local previewing. The mode `local = TRUE` is primarily for `blogdown::serve_site()` to serve the website locally. There are three major differences between `local = FALSE` and `TRUE`. When `local = TRUE`:

- The `baseurl` option in `config.toml` is temporarily overridden by `"/"` even if you have set it to a full URL like `"http://www.example.com/"`.[3] This is because when a website is to be previewed locally, links should refer to local files. For example, `/about/index.html` should be used instead of the full link `http://www.example.com/about/index.html`; the function `serve_site()` knows that `/about/index.html` means the file under the `public/` directory, and can fetch it and display the content to you, otherwise your browser will take you to the website `http://www.example.com` instead of displaying a local file.

- Draft and future posts are always rendered when `local = TRUE`, but not when `local = FALSE`. This is for you to preview draft and future posts locally. If you know the Hugo command line,[4] it means the `hugo` command is called with the flags `-D -F`, or equivalently, `--buildDrafts --buildFuture`.

- There is a caching mechanism to speed up building your website: an Rmd file will not be recompiled when its `*.html` output file is newer (in terms of file modification time). If you want to force `build_site(local = TRUE)` to recompile the Rmd file even if it is older than the HTML output, you need to delete the HTML output, or edit the Rmd file so that its modification time will be newer. This caching mechanism does not apply to

---

[3] If your `baseurl` contains a subdirectory, it will be overridden by the subdirectory name. For example, for `baseurl = "http://www.example.com/project/"`, `build_site(local = TRUE)` will temporarily remove the domain name and only use the value `/project/`.

[4] `https://gohugo.io/commands/hugo/`

`local = FALSE`, i.e., `build_site(local = FALSE)` will always recompile all Rmd files, because when you want to publish a site, you may need to recompile everything to make sure the site is fully regenerated. If you have time-consuming code chunks in any Rmd files, you have to use either of these methods to save time:

- Turn on **knitr**'s caching for time-consuming code chunks, i.e., the chunk option `cache = TRUE`.

- Do not call `build_site()`, but `blogdown::hugo_build()` instead. The latter does not compile any Rmd files, but simply runs the `hugo` command to build the site. Please use this method only if you are sure that your Rmd files do not need to be recompiled.

You do not need to worry about these details if your website is automatically generated from source via a service like Netlify, which will make use of `baseurl` and not use `-D -F` by default. If you manually publish the `public/` folder, you need to be more careful:

• If your website does not work without the full `baseurl`, or you do not want the draft or future posts to be published, you should not publish the `public/` directory generated by `serve_site()`. Always run `blogdown::build_site()` or `blogdown::hugo_build()` before you upload this directory to a web server.

• If your drafts and future posts contain (time-)sensitive information, you are strongly recommended to delete the `/public/` directory before you rebuild the site for publishing every time, because Hugo never deletes it, and your sensitive information may be rendered by a certain `build_site(local = TRUE)` call last time and left in the directory. If the website is really important, and you need to make sure you absolutely will not screw up anything every time you publish it, put the `/public/` directory under version control, so you have a chance to see which files were changed before you publish the new site.

D.4   Functions in the blogdown package

There are about 20 exported functions in the **blogdown** package, and many more non-exported functions. Exported functions are documented and you can use them after `library(blogdown)` (or via `blogdown::`). Non-exported functions are not documented, but you can access them via `blogdown:::` (the triple-colon syntax). This package is not very complicated, and consists of only about 1800 lines of R code (the number is given by the word-counting command `wc`):

```
wc -l ../R/*.R ../inst/scripts/*.R
```

```
  50 ../R/clean.R
  55 ../R/format.R
 355 ../R/hugo.R
 169 ../R/install.R
  33 ../R/package.R
 160 ../R/render.R
 171 ../R/serve.R
  28 ../R/site.R
 604 ../R/utils.R
  79 ../inst/scripts/new_post.R
  31 ../inst/scripts/render_page.R
   6 ../inst/scripts/render_rmarkdown.R
  69 ../inst/scripts/update_meta.R
1810 total
```

You may take a look at the source code (`https://github.com/rstudio/blogdown`) if you want to know more about a non-exported function. In this section, we selectively list some exported and non-exported functions in the package for your reference.

D.4.1 Exported functions

Installation: You can install Hugo with `install_hugo()`, update Hugo with `update_hugo()`, and install a Hugo theme with `install_theme()`.

Wrappers of Hugo commands: `hugo_cmd()` is a general wrapper of `system2('hugo', ...)`, and all later functions execute specific Hugo commands based on this general wrapper function; `hugo_version()` executes the command `hugo version` (i.e., `system2('hugo', 'version')` in R); `hugo_build()` executes `hugo` with optional parameters; `new_site()` executes `hugo new site`; `new_content()` executes `hugo new` to create a new content file, and `new_post()` is a wrapper based on `new_content()` to create a new blog post with appropriate YAML metadata and filename; `hugo_convert()` executes `hugo convert`; `hugo_server()` executes `hugo server`.

Output format: `html_page()` is the only R Markdown output format function in the package. It inherits from `bookdown::html_document2()`, which in turn inherits from `rmarkdown::html_document()`. You need to read the documentation of these two functions to know the possible arguments. Section 1.5 has more detailed information about it.

Helper functions: `shortcode()` is a helper function to write a Hugo shortcode `{{% %}}` in an Rmd post; `shortcode_html()` writes out `{{< >}}`.

Serving a site: `serve_site()` starts a local web server to build and preview a site continuously; you can stop the server via `stop_server()`, or restart your R session.

Dealing with YAML metadata: `find_yaml()` can be used to find content files that contain a specified YAML field with specified values; `find_tags()` and `find_categories()` are wrapper functions based on `find_yaml()` to match specific tags and categories in content files, respectively; `count_yaml()` can be used to calculate the frequencies of specified fields.

D.4.2 Non-exported functions

Some functions are not exported in this package because average users are unlikely to use them directly, and we list a subset of them below:

- You can find the path to the Hugo executable via `blogdown:::find_hugo()`.

If the executable can be found via the PATH environment variable, it just returns 'hugo'.

- The helper function modify_yaml() can be used to modify the YAML metadata of a file. It has a ... argument that takes arbitrary YAML fields, e.g., blogdown:::modify_yaml('foo.md', author = 'Frida Gomam', date = '2015-07-23') will change the author field in the file foo.md to Frida Gomam, and date to 2015-07-23. We have shown the advanced usage of this function in Section 4.1.

- We have also mentioned a series of functions to clean up Markdown posts in Section 4.1. They include process_file(), remove_extra_empty_lines(), process_bare_urls(), normalize_chars(), remove_highlight_tags(), and fix_img_tags().

- In Section D.3, we mentioned a caching mechanism based on the file modification time. It is implemented in blogdown:::require_rebuild(), which takes two arguments of filenames. The first file is the output file, and the second is the source file. When the source file is older than the output file, or the output file does not exist or is empty, this function returns TRUE.

- The function blogdown:::Rscript() is a wrapper function to execute the command Rscript, which basically means to execute an R script in a new R session. We mentioned this function in Chapter 5.

D.5   Paths of figures and other dependencies

One of the most challenging tasks in developing the **blogdown** package is to properly handle dependency files of web pages. If all pages of a website were plain text without dependencies like images or JavaScript libraries, it would be much easier for me to develop the **blogdown** package.

After **blogdown** compiles each Rmd document to HTML, it will try to detect the dependencies (if there are any) from the HTML source and copy them to the static/ folder, so that Hugo will copy them to public/ later. The detection depends on the paths of dependencies. By default, all dependencies, like R plots and libraries for HTML widgets, are generated to the foo_files/

directory if the Rmd is named `foo.Rmd`. Specifically, R plots are generated to `foo_files/figure-html/` and the rest of files under `foo_files/` are typically from HTML widgets.

R plots under `content/*/foo_files/figure-html/` are copied to `static/*/foo_files/figure-html/`, and the paths in HTML tags like `<img src="foo_files/figure-html/bar.png" />` are substituted with `/*/foo_files/figure-html/bar.png`. Note the leading slash indicates the root directory of the published website, and the substitution works because Hugo will copy `*/foo_files/figure-html/` from `static/` to `public/`.

Any other files under `foo_files/` are treated as dependency files of HTML widgets, and will be copied to `static/rmarkdown-libs/`. The original paths in HTML will also be substituted accordingly, e.g., from `<script src="foo_files/jquery/jquery.min.js">` to `<script src="/rmarkdown-libs/jquery/jquery.min.js">`. It does not matter whether these files are generated by HTML widgets or not. The links on the published website will be correct and typically hidden from the readers of the pages.[5]

You should not modify the **knitr** chunk option `fig.path` or `cache.path` unless the above process is completely clear to you, and you want to handle dependencies by yourself.

In the rare cases when **blogdown** fails to detect and copy some of your dependencies (e.g., you used a fairly sophisticated HTML widget package that writes out files to custom paths), you have two possible choices:

- Do not ignore `_files$` in the option `ignoreFiles` in `config.toml`, do not customize the `permalinks` option, and set the option `uglyURLs` to `true`. This way, **blogdown** will not substitute paths that it cannot recognize, and Hugo will copy these files to `public/`. The relative file locations of the `*.html` file and its dependencies will remain the same when they are copied to `public/`, so all links will continue to work.

- If you choose to ignore `_files$` or have customized the `permalinks` option, you need to make sure **blogdown** can recognize the dependencies. One approach is to use the path returned by the helper function

---

[5] For example, a reader will not see the `<script>` tag on a page, so it does not really matter what its `src` attribute looks like as long as it is a path that actually exists.

`blogdown::dep_path()` to write out additional dependency files. Basically this function returns the current `fig.path` option in **knitr**, which defaults to `*_files/figure-html/`. For example, you can generate a plot manually under `dep_path()`, and **blogdown** will process it automatically (copy the file and substitute the image path accordingly).

If you do not understand all these technical details, we recommend that you use the first choice, and you will have to sacrifice custom permanent links and clean URLs (e.g., `/about.html` instead of `/about/`). With this choice, you can also customize the `fig.path` option for code chunks if you want.

D.6   HTML widgets

We do not recommend that you use different HTML widgets from many R packages on the same page, because it is likely to cause conflicts in JavaScript. For example, if your theme uses the jQuery library, it may conflict with the jQuery library used by a certain HTML widget. In this case, you can conditionally load the theme's jQuery library by setting a parameter in the YAML metadata of your post and revising the Hugo template that loads jQuery. Below is the example code to load jQuery conditionally in a Hugo template:

```
{{ if not .Params.exclude_jquery}}
<script src="path/to/jquery.js"></script>
{{ end }}
```

Then if you set `exclude_jquery: true` in the YAML metadata of a post, the theme's jQuery will not be loaded, so there will not be conflicts when your HTML widgets also depend on jQuery.

Another solution is the **widgetframe** package[6] (Karambelkar, 2017). It solves this problem by embedding HTML widgets in `<iframe></iframe>`. Since an

---

[6]https://github.com/bhaskarvk/widgetframe

iframe is isolated from the main web page on which it is embedded, there will not be any JavaScript conflicts.

A widget is typically not of full width on the page. To set its width to 100%, you can use the chunk option `out.width = "100%"`.

D.7 Version control

If your website source files are under version control, we recommend that you add at least these two folder names to your `.gitignore` file:

```
blogdown
public
```

The `blogdown/` directory is used to store cache files, and they are most likely to be useless to the published website. Only **knitr** may use them, and the published website will not depend on these files.

The `public/` directory should be ignored if your website is to going to be automatically (re)built on a remote server such as Netlify.

As we mentioned in Section D.5, R plots will be copied to `static/`, so you may see new files in GIT after you render an Rmd file that has graphics output. You need to add and commit these new files in GIT, because the website will use them.

Although it is not relevant to **blogdown**, macOS users should remember to ignore `.DS_Store` and Windows users should ignore `Thumbs.db`.

If you are relatively familiar with GIT, there is a special technique that may be useful for you to manage Hugo themes, which is called "GIT submodules." A submodule in GIT allows you to manage a particular folder of the main repository using a different remote repository. For example, if you used the default `hugo-lithium-theme` from my GitHub repository, you might want to sync it with my repository occasionally, because I may update it from time to time. You can add the GIT submodule via the command line:

```
git submodule add \
  https://github.com/yihui/hugo-lithium-theme.git \
  themes/hugo-lithium-theme
```

If the folder themes/hugo-lithium-theme exists, you need to delete it before
adding the submodule. Then you can see a SHA string associated with the
"folder" themes/hugo-lithium-theme in the GIT status of your main repository indicating the version of the submodule. Note that you will only see the
SHA string instead of the full content of the folder. Next time when you want
to sync with my repository, you may run the command:

```
git submodule update --recursive --remote
```

In general, if you are happy with how your website looks, you do not need to
manage the theme using GIT submodules. Future updates in the upstream
repository may not really be what you want. In that case, a physical and fixed
copy of the theme is more appropriate for you.

D.8   The default HTML template

As we mentioned in Section 1.5, the default output format for an Rmd document in **blogdown** is blogdown::html_page. This format passes a minimal
HTML template to Pandoc by default:

```
<!-- BLOGDOWN-HEAD -->
$for(header-includes)$
$header-includes$
$endfor$
$if(highlighting-css)$
<style type="text/css">
$highlighting-css$
</style>
$endif$
```

```
$for(css)$
  <link rel="stylesheet" href="$css$" type="text/css" />
$endfor$
<!-- /BLOGDOWN-HEAD -->

$for(include-before)$
$include-before$
$endfor$
$if(toc)$
<div id="$idprefix$TOC">
$toc$
</div>
$endif$

$body$

$for(include-after)$
$include-after$
$endfor$
```

You can find this template file via `blogdown:::pkg_file('resources', 'template-minimal.html')` in R, and this file path is the default value of the `template` argument of `html_page()`. You may change this default template, but you should understand what this template is supposed to do first.

If you are familiar with Pandoc templates, you should realize that this is not a complete HTML template, e.g., it does not have the tags `<html>`, `<head>`, or `<body>`. That is because we do not need or want Pandoc to return a full HTML document to us. The main thing we want Pandoc to do is to convert our Markdown document to HTML, and give us the body of the HTML document, which is in the template variable `$body$`. Once we have the body, we can further pass it to Hugo, and Hugo will use its own template to embed the body and generate the full HTML document. Let's explain this by a minimal example. Suppose we have an R Markdown document `foo.Rmd`:

```
---
title: "Hello World"
author: "Yihui Xie"
---


I found a package named **blogdown**.
```

It is first converted to an HTML file `foo.html` through `html_page()`, and note that YAML metadata are ignored for now:

```
<!-- BLOGDOWN-HEAD -->
<!-- /BLOGDOWN-HEAD -->


I found a package named <strong>blogdown</strong>.
```

Then **blogdown** will read the YAML metadata of the Rmd source file, and insert the metadata into the HTML file so it becomes:

```
---
title: "Hello World"
author: "Yihui Xie"
---


I found a package named <strong>blogdown</strong>.
```

This is the file to be picked up by Hugo and eventually converted to an HTML page of a website. Since the Markdown body has been processed to HTML by Pandoc, Hugo will basically use the HTML. That is how we bypass Hugo's Markdown engine BlackFriday.

Besides `$body$`, you may have noticed other Pandoc template variables like `$header-includes$`, `$css$`, `$include-before$`, `$toc$`, and `$include-after$`. These variables make it possible to customize the `html_page` format. For example, if you want to generate a table of contents, and apply an additional CSS stylesheet to a certain page, you may set `toc` to `true` and pass the stylesheet path to the `css` argument of `html_page()`, e.g.,

```
---
title: "Hello World"
author: "Yihui Xie"
output:
  blogdown::html_page:
    toc: true
    css: "/css/my-style.css"
---
```

There is also a pair of HTML comments in the template: `<!--` `BLOGDOWN-HEAD -->` and `<!-- /BLOGDOWN-HEAD -->`. This is mainly for `method = 'html_encoded'` in `blogdown::build_site()` (see Section D.9). This pair of comments is used to mark the HTML code fragment that should be moved to the `<head>` tag of the final HTML page. Typically this code fragment contains links to CSS and JavaScript files, e.g., those requested by the user via the `css` argument of `html_page()`, or automatically generated when HTML widgets are used in an Rmd document. For `method = 'html'`, this code fragment is not moved, which is why the final HTML page may not conform to W3C standards. If you want to customize the template, you are recommended to use this pair of comments to mark the HTML code fragment that belongs to the `<head>` tag.

D.9 Different building methods

If your website does not contain any Rmd files, it is very straightforward to render it — just a system call to the `hugo` command. When your website contains Rmd files, **blogdown** has provided two rendering methods to compile these Rmd files. A website can be built using the function `blogdown::build_site()`:

```
build_site(local = FALSE, method = c("html", "custom"),
  run_hugo = TRUE)
```

As mentioned in Section 1.4, the default value of the `method` argument is determined by the global option `blogdown.method`, and you can set this option in `.Rprofile`.

For `method = 'html'`, `build_site()` renders `*.Rmd` to `*.html`, and `*.Rmarkdown` to `*.markdown`, and keeps the `*.html/*.markdown` output files under the same directory as `*.Rmd/*.Rmarkdown` files.

An Rmd file may generate two directories for figures (`*_files/`) and cache (`*_cache/`), respectively, if you have plot output or HTML widgets (Vaidyanathan et al., 2017) in your R code chunks, or enabled the chunk option `cache = TRUE` for caching. In the figure directory, there will be a subdirectory `figure-html/` that contains your plot output files, and possibly other subdirectories containing HTML dependencies from HTML widgets (e.g., `jquery/`). The figure directory is moved to `/static/`, and the cache directory is moved to `/blogdown/`.

After you run `build_site()`, your website is ready to be compiled by Hugo. This gives you the freedom to use deploying services like Netlify (Chapter 3), where neither R nor **blogdown** is available, but Hugo is.

For `method = 'custom'`, `build_site()` will not process any R Markdown files, nor will it call Hugo to build the site. No matter which method you choose to use, `build_site()` will always look for an R script `/R/build.R` and execute it if it exists. This gives you the complete freedom to do anything you want for the website. For example, you can call `knitr::knit()` to compile Rmd to Markdown (`*.md`) in this R script instead of using `rmarkdown::render()`. This feature is designed for advanced users who are really familiar with the **knitr** package[7] and Hugo or other static website generators (see Chapter 5).

---

[7] Honestly, it was originally designed for Yihui himself to build his own website, but he realized this feature could actually free users from Hugo. For example, it is possible to use Jekyll (another popular static site generator) with **blogdown**, too.

E

Personal Experience

I started blogging at blogchina.com in 2005, moved to blog.com.cn, then MSN Space, and finally purchased my own domain `yihui.name` and a virtual host. I first used a PHP application named Bo-Blog, then switched to WordPress, and then Jekyll. Finally I moved to Hugo. Although I have moved several times, all my posts have been preserved, and you can still see my first post in Chinese in 2005. I often try my best not to introduce broken links (which lead to the 404 page) every time I change the backend of my website. When it is too hard to preserve the original links of certain pages, I will redirect the broken URLs to the new URLs. That is why it is important for your system to support redirections, and in particular, 301 redirections (Netlify does a nice job here). Here are some of my redirection rules: `https://github.com/rbind/yihui/blob/master/static/_redirects`. For example, `http://yihui.name/en/feed/` was the RSS feed of my old WordPress and Jekyll blogs in English, and Hugo generates the RSS feed to `/en/index.xml` instead, so I need to redirect `/en/feed/` to `/en/index.xml`.

Google has provided several tools to help you know more information about your website. For example, Google Analytics[1] can collect visitor statistics and give speed suggestions for your website. Google Webmasters[2] can show you the broken links it finds. I use these tools frequently by myself.

I firmly believe in the value of writing. Over the years, I have written more than 1000 posts in Chinese and English. Some are long, and most are short. The total size of these text files is about 5 Mb. In retrospect, most posts are probably not valuable to general readers (some are random thoughts, and some are my rants), but I feel I benefitted a lot from writing in two aspects:

1. If I sit down and focus on writing a small topic for a while, I of-

---

[1] `https://analytics.google.com`
[2] `https://www.google.com/webmasters/`

ten feel my thoughts will become clearer. A major difference between writing and talking is that you can always reorganize things
and revise them when writing. I do not think writing on social me-
dia counts. 140 characters may well be thoughtful, but I feel there
is so much chaos there. It is hard to lay out systematic thoughts
only through short messages, and these quick messages are often
quickly forgotten.

2.   I know some bloggers are very much against comments, so they
     do not open comments to the public. I have not had a very nega-
     tive experience with comments yet. On the contrary, I constantly
     find inspirations from comments. For example, I was thinking[3]
     if it was possible to automatically check R packages on the cloud
     through Travis CI. At that time (April 2013), I believe not many peo-
     ple in the R community had started using Travis CI, although I'm
     not sure if I was the first person experimenting with this idea. I
     felt Travis CI could be promising, but it did not support R back
     then. Someone named Vincent Arel-Bundock (I still do not know
     him) told me a hack in a comment, which suddenly lit up my mind
     and I quickly figured out a solution. In October 2013, Craig Citro
     started more solid work on the R support on Travis CI. I do not
     know if he saw my blog post. Anyway, I think Travis CI has made
     substantial impact on R package developers, which is a great thing
     for the R community.

Yet another relatively small benefit is that I often go to my own posts to learn
some technical stuff that I have forgotten. For example, I find it difficult to
remember the syntax of different types of zero-width assertions in Perl-like
regular expressions: `(?=...)`, `(?!...)`, `(?<=...)`, and `(?<!...)`. So I wrote a
short blog post and gave myself a few minimal examples. After going back
to that post a few times, finally I can remember how to use these regular
expressions.

---

[3]`https://yihui.name/en/2013/04/travis-ci-for-r/`

Bibliography

Allaire, J., McPherson, J., Xie, Y., Luraschi, J., Ushey, K., Atkins, A., Wickham, H., Cheng, J., and Chang, W. (2017). *rmarkdown: Dynamic Documents for R*. R package version 1.7.

Cheng, J. (2017). *later: Utilities for Delaying Function Execution*. R package version 0.6.

Csárdi, G. (2017). *processx: Execute and Control System Processes*. R package version 2.0.0.1.

Karambelkar, B. (2017). *widgetframe: Htmlwidgets in Responsive iframes*. R package version 0.3.0.

R Core Team (2017). *R: A Language and Environment for Statistical Computing*. R Foundation for Statistical Computing, Vienna, Austria.

Vaidyanathan, R., Xie, Y., Allaire, J., Cheng, J., and Russell, K. (2017). *htmlwidgets: HTML Widgets for R*. R package version 0.9.

Wickham, H. (2017). *pkgdown: Make Static HTML Documentation for a Package*. http://hadley.github.io/pkgdown, https://github.com/hadley/pkgdown.

Xie, Y. (2016). *bookdown: Authoring Books and Technical Documents with R Markdown*. Chapman and Hall/CRC, Boca Raton, Florida. ISBN 978-1138700109.

Xie, Y. (2017a). *animation: A Gallery of Animations in Statistics and Utilities to Create Animations*. R package version 2.5.

Xie, Y. (2017b). *bookdown: Authoring Books and Technical Documents with R Markdown*. R package version 0.6.

Xie, Y. (2017c). *knitr: A General-Purpose Package for Dynamic Report Generation in R*. R package version 1.18.

Xie, Y. (2017d). *servr: A Simple HTTP Server to Serve Static Files or Dynamic Documents*. R package version 0.8.

Xie, Y. (2017e). *xaringan: Presentation Ninja*. R package version 0.4.3.

Index